To Tee ♡



[illegible]

# Last Stand

# LAST STAND

## Protected Areas and the Defense of Tropical Biodiversity

Edited by
Randall Kramer
Carel van Schaik
Julie Johnson

*New York Oxford*
OXFORD UNIVERSITY PRESS
*1997*

Oxford University Press

Oxford New York
Athens Auckland Bangkok Bogota Bombay Buenos Aires
Calcutta Cape Town Dar es Salaam Delhi Florence Hong Kong
Istanbul Karachi Kuala Lumpur Madras Madrid Melbourne
Mexico City Nairobi Paris Singapore Taipei Tokyo Toronto

and associated companies in
Berlin Ibadan



Published by Oxford University Press, Inc.
198 Madison Avenue, New York, New York 10016



Library of Congress Cataloging-in-Publication Data
Last stand : protected areas and the defense of tropical biodiversity
/ edited by Randall Kramer, Carel van Schaik, Julie Johnson.
p. cm.
Includes bibliographical references and index.
ISBN 0-19-509554-5
1. Biological diversity conservation–Tropics. I. Kramer,
Randall A. II. Schaik, Carel van. III. Johnson, Julie.
QH75.L365 1997 96-22337
333.95'16'0913–dc20

9 8 7 6 5 4 3 2 1

Printed in the United States of America
on acid free paper

To

Lydia and Hannah
Jaap and Anna
Anne

We hope you will experience the wonder your parents have known in the rare tracts of rain forest that still exist.

# Preface

This book is the culmination of a several-year effort to bring together people from a variety of disciplines, national backgrounds, and work environments to share ideas about the conservation of biodiversity in the tropics.

In 1992, the Center for Tropical Conservation at Duke University, with funding from the U.S. Agency for International Development, supported our first efforts to examine issues in biodiversity conservation that eluded solution within traditional academic boundaries. Priya Shyamsundar and Nick Salafsky provided early input that helped shape our overall approach, as we initially brought together the perspectives of the ecologist and the economist. A series of seminars grew out of these early collaborations. The resulting lively discussions among faculty, students, and visitors to Duke encouraged us to organize an international workshop in December 1993 to explore realistic ideas that might increase the chances of saving some of the fast-disappearing species and habitats in tropical ecosystems. Funding for the workshop was generously provided by The Howard Gilman Foundation.

We thank the workshop participants who shared their ideas and time, and critically shaped the chapters contained in this volume: Kiran Asher, Katrina Brandon, John Browder, Norm Christensen, Jason Clay, Gilbert Isabirye-Basuta, Bruce Larson, Owen Lynch, Kathy MacKinnon, Hemmo Muntingh, Kate Newman, Carlos Peres, Peter Principe, Tahir Qadri, Kent Redford, John Robinson, Nick Salafsky, Steve Sanderson, Kathryn Saterson, Narendra Sharma, Priya Shyamsundar, Tom Struhsaker, Bill Sugrue, Yucatan Teixeira da Silva, John Terborgh, and Jan Wind.

We thank Jim Smith, Marcello Guidi, and Michael Gary of The Howard Gilman Foundation, and the staff of White Oak Plantation for providing an unsurpassed workshop environment. Lisa Davenport, Madhu Rao, Danny Rothberg, and Maxine Sanders provided expert logistical support for the workshop. John Huyler and Martha Tableman of the Keystone Foundation ably facilitated the workshop.

We also thank Norm Christensen, Malcolm Gillis, and John Terborgh for providing advice and encouragment. In his own way, each believes that science can and must inform natural resource policy, and this view pervades the book.

We especially thank the authors for their hard work and patience with the editors, and their willingness to help shape this collection of papers. Their academic disciplines, their different experiences of the tropics, and their individual voices are all reflected in these chapters, yet—hearteningly—many common themes and messages emerge. Finally, Tom Burroughs provided exemplary technical editing, and helped us to express those themes and messages more clearly.

# Contents

Contributors xi

1. Preservation Paradigms and Tropical Rain Forests 3
   *Randall A. Kramer and Carel P. van Schaik*
2. Minimizing Species Loss: The Imperative of Protection 15
   *John Terborgh and Carel P. van Schaik*
3. The Ecological Foundations of Biodiversity Protection 36
   *Kathy MacKinnon*
4. The Silent Crisis: The State of Rain Forest Nature Preserves 64
   *Carel P. van Schaik, John Terborgh, and Barbara Dugelby*
5. Policy and Practical Considerations in Land-Use Strategies for Biodiversity Conservation 90
   *Katrina Brandon*
6. Biodiversity Politics and the Contest for Ownership of the World's Biota 115
   *Steven E. Sanderson and Kent H. Redford*
7. User Rights and Biodiversity Conservation 133
   *Marie Lynn Miranda and Sharon LaPalme*
8. Tropical Forest Biodiversity Protection: Who Pays and Why 162
   *Randall A. Kramer and Narendra Sharma*
9. Compensation and Economic Incentives: Reducing Pressure on Protected Areas 187
   *Paul J. Ferraro and Randall A. Kramer*
10. Toward a New Protection Paradigm 212
    *Carel P. van Schaik and Randall A. Kramer*

Index 231

# Contributors

*Katrina Brandon* is a Senior Fellow with the Parks in Peril Program of The Nature Conservancy, adjunct faculty in the Conservation Biology and Sustainable Development Program at the University of Maryland, and a consultant on conservation and development issues. Dr. Brandon is currently coediting a book analyzing the influence of social and political factors at protected area sites in Latin America and the Caribbean. She holds an M.A. in Inter-American studies from the University of Miami and M.S. and Ph.D. degrees in development sociology and planning from Cornell University.

*Barbara Dugelby* is Conservation Scientist with The Nature Conservancy, Latin America and Caribbean Division. She received her Ph.D. in tropical ecology and conservation from Duke University's Nicholas School of the Environment. She served as Instructor in the School of the Environment and Administrative Coordinator for the Center for Tropical Conservation at Duke. Her research interests lie in tropical forest ecology and conservation, protected area management, and community-based resource management. Since 1986, Dr. Dugelby has conducted or assisted with research in Panama, Guatemala, Madagascar, Belize, Indonesia, and Peru. Between 1992 and 1993, she served as Conservation International's Non-timber Research Coordinator in the Maya Biosphere Reserve.

*Paul J. Ferraro* received a B.A. in biology and history and his M.S. in resource economics from Duke University's Nicholas School of the Environment. He has worked extensively in Madagascar on developing local-level incentives for biodiversity conservation and on assessing the local impacts of protected area policies. Currently he is an independent consultant working on conservation and natural resource use issues.

*Julie A. Johnson* is Assistant Research Professor at Duke University's Nicholas School of the Environment and Assistant Director for Special Programs in the Center for International Development Research at the Terry Sanford Institute for Public Policy. She holds a Ph.D. in zoology from the University of Edinburgh and spent a postdoctoral fellowship studying health policy and environmental health at the U.S. Agency for International Development. Dr. Johnson's research interests are in primate social behavior and the effects in general of social behavior and relationships on reproductive success and, ultimately, evolutionary processes. Fieldwork in Tanzania and Kenya has led her to concentrate recently on the conflicting human interests shaping natural resource use.

*Randall A. Kramer* is an Associate Professor at Duke University's Nicholas School of the Environment and in the Department of Economics. He received his Ph.D. at the University of California, Davis, and taught at Virginia Polytechnic Institute and State University before coming to Duke. Dr. Kramer has been a visiting fellow at the Yale Economic Growth Center. His current research areas are environmental valuation of natural environments and economic analysis of wetlands and biodiversity policies. Dr. Kramer has served as a consultant on biodiversity issues with the World Bank, Asian Development Bank, and the U.S. Agency for International Development in Madagascar, Indonesia, and Sri Lanka.

*Sharon LaPalme* is a research assistant at Duke University's Sanford Institute of Public Policy. She holds master's degrees from the Sanford Institute and Duke University's Nicholas School of the Environment. Her research has included U.S. and tropical forest resource policy, and forest ecology. Her other interests and activities include sustainable agriculture and international environmental issues.

*Kathy MacKinnon* was born in the United Kingdom and holds a Ph.D. in ecology from the University of Oxford. She first experienced tropical forests in Sumatra in 1971 and subsequently spent 11 years in the Asian tropics, mainly Indonesia, working on research, conservation and environmental projects, with governmental agencies, nongovernment organizations, and universities. Coauthor of several conservation texts including *Managing Protected Areas in the Tropics* and the *IUCN Reviews of the Protected Areas of the Indomalayan and Afrotropical Realms*, Dr. MacKinnon has also assisted with preparations of the Indonesian Biodiversity Action Plan. She has written extensively on wildlife and ecology in Asia, including a textbook and popular books. She is currently employed as a biodiversity specialist at the World Bank in Washington, D.C., providing technical support to biodiversity projects funded through the Global Environmental Facility.

*Marie Lynn Miranda* is an Assistant Professor at the Nicholas School of the Environment at Duke University. She received her Ph.D. in economics from Harvard University. Her areas of specialization are natural resource and environmental economics and policy. Prior to receiving her doctorate, she worked for both the National Marine Fisheries Service and the U.S. Forest Service. Her current research is split between international and domestic environmental policy issues. Internationally, Dr. Miranda works primarily on tropical forestry issues, with country experience in Nepal, Indonesia, Honduras, Costa Rica, and Malaysia. Domestically, she focuses on land use planning issues, including agriculture-forestry-development tradeoffs and the management of municipal solid waste.

*Kent H. Redford* is the Director for Conservation Science and Stewardship of the Latin America and Caribbean Division of The Nature Conservancy. He received his Ph.D. in biology at Harvard University and has taught and served as the Director of the Program in Tropical Conservation at the University of Florida. His current research areas include the interactions between natural resource use by traditional peoples and biodiversity conservation, the role of mammals in neotropical forests, and the politics of biodiversity. Dr. Redford has served as a consultant with the World Bank, the World Wildlife Fund, and Conservation International.

*Steven E. Sanderson* is Professor of Political Science and Co-Director of the Tropical Conservation and Development Program in the Center for Latin American Studies at the University of Florida. He has studied the politics of rural poverty and natural resource use throughout Latin America, with special attention to Mexico and Brazil. He served as Ford Foundation Program Officer for Rural Poverty and Resources in Brazil and led the design of the foundation's Amazon program. Dr. Sanderson is a member of the National Academy of Sciences Committee on the Human Dimensions of Global Environmental Change. His research includes the impact of international trade on agricultural land use, rural incomes policies, domestic food policies, and conversion of natural cover to human use. He is also conducting research on the role of property rights in biodiversity conservation and the impact of institutional changes on land use patterns among the poor. Dr. Sanderson's most recent book is *The Politics of Trade in Latin American Development*.

*Narendra Sharma* is a principal economist at the World Bank in Washington, D.C., and is the primary author of the Bank's forest policy. He received a Ph.D. in agricultural economics and economic development from Virginia Polytechnic Institute and State University. His research interests are in applied economics, project design, and policy analyses. Dr. Sharma has worked in developing countries on policy issues related to conservation and sustainable development, poverty, natural resource management, and policy dialogue. His current research focuses on quantification of environmental impacts and local participation.

*John Terborgh* is a James B. Duke Professor in the Nicholas School of the Environment at Duke University. He received his Ph.D. in biology at Harvard University and taught at Princeton University before coming to Duke. His interests lie in the fields of tropical ecology and conservation. At different times in his career Dr. Terborgh has studied birds, primates, herbs, and forest trees, and has directed student projects involving butterflies, lizards, amphibians, and crocodilians. The common denominator in all this work has been the goal of solving problems of general ecological interest using a comparative approach. Some typical comparisons have involved seasonal patterns in resource utilization by forest primates, habitat use by Amazonian birds, and latitudinal variation in the structure of mature forests. Applications of ecology to conservation have increasingly become a central theme of his work. Dr. Terborgh regards as particularly important the need to understand the many consequences of habitat fragmentation, especially those related to the disruption of trophic level processes.

*Carel P. van Schaik* is a professor at Duke University's Departments of Biological Anthropology and Anatomy, and Zoology, and at the Nicholas School of the Environment. He received a Ph.D. in behavioral ecology from Utrecht University in the Netherlands and was a postdoctoral fellow at Princeton University and a senior research fellow at the Royal Netherlands Academy of Arts and Sciences. His research focuses on seasonality and frugivory in tropical rain forests and on primate social evolution and social dynamics. Dr. van Schaik's study of macaques and orangutans in Sumatra, Indonesia, has spanned more than twenty years.

Last Stand

# 1

# Preservation Paradigms and Tropical Rain Forests

*Randall A. Kramer and Carel P. van Schaik*

Tropical rain forests are disappearing rapidly as a result of increasing human encroachment. During the past century, tropical rain forests have been reduced to about half of their original area. And the rate of deforestation is accelerating, fueled by population growth in developing countries and resource demands in the developed countries. The remaining forests are subject to increasingly intensive human use. Deforestation, fragmentation, and exploitation cause a plethora of problems, including soil erosion; siltation of rivers, lakes, and estuaries; increased flooding and droughts; release of carbon dioxide and other greenhouse gases into the atmosphere; and loss of species. In recent years, these problems have become the subject of international concern. This book focuses on the loss of biodiversity in tropical rain forests and on the role of protected areas in stemming the loss.

This chapter examines the meaning of biodiversity and the history of the park movement in the tropics. What began as protection of habitat through the exclusion of people has transformed into sustainable use of biological resources. This new emphasis provides local control of important resources and greater income, but does it conserve habitat and species? We will argue that a renewed focus on protected areas as the primary storehouse of biodiversity is needed. We will also make the case for a focus on the tropical rain forest biome and will conclude with an overview of the rest of the book.

## INTERPRETATIONS OF BIODIVERSITY

In its strict sense, biodiversity refers to the "variety and variability among living organisms and the ecological complexes in which they occur" (Office of Technology Assessment, U.S. Congress, 1987:3). This definition can be extended both downward to cover genetic variability within a species and upward to include habitat and ecosystem diversity. In practical terms, however, biodiversity is most profitably expressed as species diversity (weighted for rarity, endemism, and taxonomic distinctiveness, if necessary) at the landscape level (see chapter 6). We adopt this definition of biodiversity.

During the past few years, attempts to link rain forest protection with sustainable development have led to a noticeable expansion of the meaning of the phrase "biodiversity conservation." In this increasingly popular view, biodiversity has come to represent ecological services and products such as clean air and water. This definition has led to a shift away from species protection and toward sustainable use.

This dilution of the biodiversity concept has been very effective in increasing the appeal of conservation efforts to a use-oriented audience. But clearly, its adoption must lead to a dramatic difference in conservation strategies. Witness the following statement in a recent major international policy document, the Global Biodiversity Strategy: "Biodiversity conservation entails a shift from a defensive posture—protecting nature from the impacts of development—to an offensive effort seeking to meet peoples' needs from biological resources while ensuring the long-term sustainability of Earth's biotic wealth" (WRI et al., 1992:5). Thus, this utilitarian perspective leads to a conservation strategy in which there is little or no need for strictly protected areas. In contrast, the species perspective, which emphasizes the ethical and scientific values of biological species regardless of their utility for humans, leads to an emphasis on protection of examples of natural ecosystems, little modified by human action.

The impact of this difference in conservation strategy on biodiversity maintenance depends on the impact of use on the survival of local species populations. Utilization may be more likely to lead to local extinction of rain forest species. The high diversity of these communities means that each species tends to be rare. Moreover, in comparison with organisms from other habitats, those of rain forests tend to have slow life histories, which translates into low productivities (Robinson and Redford, 1991). Thus, ironically, adherence to a diluted biodiversity concept may contribute in some cases to a reduction of biodiversity (Robinson, 1993). Such limited losses in biodiversity are, however, unlikely to compromise dramatically the ecosystem's ability to provide ecological services or its overall productivity, because there is usually some redundancy in the roles of species in ecological communities (Schulze and Mooney, 1993).

The upshot of this new emphasis on sustainability rather than protection has been a reduced commitment to protected areas. In addition, two other issues have become intertwined with the increased emphasis on sustainability and have helped reduce the commitment to protected areas: a change in view of the role of parks, and a heightened interest in the plight of indigenous people.

## CHANGING VIEWS OF PROTECTED AREAS

There is considerable evidence to show that many regions sustained heavy loss of species, particularly large birds and mammals, after humans first colonized them (e.g., Wilson, 1992). After this first wave of extinctions, however, many societies developed a stronger conservation ethic, establishing sacred groves, taboos, and hunting preserves that left certain amounts of habitat relatively untouched (MacKinnon et al., 1986; Primack, 1993).

The birth of the modern conservation movement began with the establishment of national parks in the United States. The first, Yellowstone National Park, was established in 1872 (Hays, 1987). In 1916, the U.S. Congress passed the Organic Act, which established the National Park Service and charged it with the following goals: "to conserve the scenery and the natural and historic objects and the wildlife therein and to provide for the enjoyment of the same by such manner and by such means as will leave them unimpaired for the enjoyment of future generations" (Public Law 64-235, 1916:535).

The concept of setting aside areas of undisturbed habitat for the enjoyment of current and future generations established a precedent of preserving pristine habitats with no exploitation of resources. The notion of keeping ecosystems functioning in their natural state in parks by excluding resident people contrasted with the park tradition in Europe. Densely populated Europe went through a major transformation of its landscape several centuries ago, and most large mammals went extinct. As a result, the European concept of a national park is strikingly different, including landscapes that have a much stronger human imprint and, indeed, require human action to be maintained in their present shape.

Most of the tropical rain forest parks planned in the past three decades have followed the U.S. model. (Of course, in the savannahs there was a much stronger emphasis on consumptive use by safaris.) However, conflicts between park managers and expanding local populations made it increasingly difficult to maintain the exclusion of people. The notion of barring people from parks began to be equated with a second wave of colonialism, and political pressure mounted in the developing countries to change the concept of conservation. In the 1970s, this dissatisfaction found its expression in the United Nations Educational, Scientific and Cultural Organization's (UNESCO) Man and the Biosphere Program, which introduced the notion of integrating conservation and development (Batisse, 1982).

In response to mounting conservation concerns, the World Conservation Strategy (IUCN et al., 1980) and its successor, Caring for the Earth (IUCN et al., 1991), defined three objectives of sustainable development: (1) the maintenance of essential ecological processes and services,

(2) the sustainable use of natural resources, and (3) the maintenance of biological diversity. Today, many international development agencies and international conservation agencies are pursuing strategies largely based on sustainable use of biological resources. This approach has been successful in alerting policymakers and project planners to the need of basing development on a sound ecological foundation.

As a result, these three objectives have become the focus of many subsequent bilateral and multilateral projects. For instance, while 31 percent of the projects funded by the U.S. Agency for International Development's (USAID) Tropical Forests and Biological Diversity Program for 1990–1991 had "protection" as one of their primary objectives, 59 percent of the projects had "community development" as a primary objective (USAID, 1992). Also, 65 percent of the projects funded in 1993 by USAID's Biodiversity Support Program, a subcategory of the larger program, are best described as "community development" projects (Peres, 1994). In sharp contrast, in the project portfolio of the Wildlife Conservation Society, a nongovernmental organization in the United States with a traditional wildlife conservation focus, only 7 percent of the projects can be categorized as having "community development" as a primary objective (WCS, 1993).

Clearly, few projects are now focusing directly on the biodiversity objective. Instead, most projects assume that satisfying the first two objectives, which are of immediate economic concern, will also adequately protect species (Redford and Sanderson, 1992). Protecting biodiversity in itself is widely perceived as a matter of less economic importance. As we argued above, however, maintaining ecological services or sustainable use of certain resources is not a guarantee that biodiversity is fully maintained, especially in rain forests (see also Gorchov, 1994).

## THE PLIGHT OF INDIGENOUS PEOPLE

Another factor contributing to the current popular focus on sustainable use has been a rising tide of interest in the plight of indigenous people (Cernea, 1991; West and Brechin, 1991). Traditional users of forests have over the years lost their rights to forest use to governments, new migrants, and commercial enterprises. Newly declared protected areas have often added to this onslaught by imposing limitations on traditional users or excluding them altogether. The new claimants for forest resources often ignored the customary practices and systems of community ownership practiced by local people (see chapter 7). This led to an international political movement to protect the rights of indigenous people.

Also, it is often claimed that forest resources would be well managed if only the traditional users were allowed to maintain control. It is, indeed, widely believed that traditional communities use their resources in a sustainable manner (McNeely, 1988). This belief is based on the fact that traditional communities lived at low densities, had limited technology, and practiced subsistence rather than commercial utilization. Unfortunately, given growing population pressure, increased access to modern technology, increasing market orientation, and steady erosion of traditional cultures, there no longer are guarantees that biodiversity objectives will be any more likely to be achieved if resource control is placed in the hands of indigenous groups.

The ascendancy of sustainable use as the dominant approach to conservation, combined with this concern for the well-being of traditional users, has prompted many people interested in improving the integrity of protected areas to focus on improving the socioeconomic status of people living nearby. Unfortunately, the projects generated by this compromise, in which buffer zones adjacent to protected areas are created for limited extractive and productive uses, have largely failed to maintain the integrity of the protected cores, although some have succeeded in improving the livelihood of the local people (Wells and Brandon, 1992) (see chapter 5).

In sum, the major processes of biodiversity erosion—the loss and fragmentation of habitat, and the overexploitation of species—continue almost unabated, around and even inside protected rain forest areas. So, while the World Conservation Strategy and other similar policy statements have made biodiversity conservation more palatable to politicians, they have also failed to address the processes underlying the rapid loss of species.

## FUNDAMENTAL CONFLICT BETWEEN CONSERVATION AND DEVELOPMENT?

Conservation and development are no doubt inextricably linked. At the national and international levels, conservation is not affordable without economic development, whereas economic development cannot be sustained without conservation. At each specific rain forest locality, however, conservation of biodiversity is usually incompatible with extractive uses, especially commercial ones, let alone with partial conversion of land and the resulting fragmentation—in sum, with most forms of economic development. Locally, therefore, there is a conflict between conservation and development.

Much of this book is devoted to reviewing the evidence that produced this thesis and to exploring its consequences for conservation strategies. This message inevitably diverges from the prevailing paradigm, which emphasizes sustainable use as the solution to the biodiversity problem. Instead, we focus on maintaining the integrity of protected areas as a key component of an overall biodiversity protection strategy. This does not make our message antidevelopment; indeed, an important means for reducing pressure on protected areas is to encourage economic development elsewhere. Our message is that we need to pursue conservation and development goals in a broader landscape context if we are to protect much of the remaining tropical forest biodiversity. Because poverty is a root cause of much of the biodiversity losses, a strong emphasis on socially equitable economic development is crucial for the success of biodiversity conservation efforts.

## THE ECONOMIC VALUE OF BIODIVERSITY

What started out as an effort to preserve natural areas for ethical and aesthetic reasons has, as we have seen, gradually adopted economic overtones. Hence, it has become popular to emphasize the economic value of tropical rain forests—for example, their value as wild relatives of important crop species or their unfulfilled potential as new sources of pharmaceuticals, industrial chemicals, and countless other useful products. Likewise, the economic benefits of the environmental services provided by intact forests have at last been recognized and are beginning to be included in value calculations. This emphasis on economic benefits evolved partly out of a desire to convince policymakers that intact forests have substantial economic values, sometimes even exceeding those generated by timber extraction or agriculture. Although this approach is a step in the right direction, it does not capture the full economic value of forests.

Nonuse values are also an important component of the total economic value of forests. Empirical evidence is emerging that nonuse values can be quite large for forested habitats and the species they contain (Kramer et al., 1992). The problem is that nonuse values are seldom measured in economic studies of forests, in part because the methods for measuring nonuse values have been developed fairly recently and because relatively few applications have occurred in tropical forest settings. Given that these nonuse values are likely to be a significant component of the economic value of biodiversity protection, leaving them out of valuation calculations can lead to serious undervaluation.

If we observe what has happened in the industrialized countries over the past several decades, we can see some reason for guarded optimism. Many tropical forest countries with growing income levels can be expected to assign greater value to their wildlands as citizens demand intact forests for recreational or intrinsic reasons. Although economic arguments that recognize use and nonuse values can be supportive of protection, decisions about protection will ultimately depend on political will. That, in turn, will depend on the level of affluence—emphasizing once again that conservation and economic development are inextricably linked.

## WHY TROPICAL FORESTS WARRANT SPECIAL ATTENTION

There is hardly a region or ecosystem on the planet that does not currently face some grave threat to its integrity or its very existence. We have chosen to focus on tropical rain forests. Focusing on one ecosystem type allows the analyses and proposed solutions to be more concrete and specific. Apart from the fact that our contributors have significant experience in the tropics, there are several powerful reasons why the problem we are addressing in this book–the incompatibility of exploitation and biodiversity preservation–is focused on this biome. The tropical rain forests have high biodiversity, high vulnerability to human disturbance, and (surprising to some) high opportunities for conservation.

Biodiversity is not uniformly distributed. By all accounts, it is greatest in the tropics. In terrestrial ecosystems, biodiversity (in the strict sense) reaches its pinnacle in the lowland evergreen forests located in moist to everwet climates, collectively known as tropical rain forests. Species richness in these forests tends to be greatest where rainfall is least seasonal (Wright, 1992). Tropical rain forests contain more than half of all known species (Wilson, 1992), and it is predicted that the great majority of the as-yet-undiscovered species reside in the tropics (WCMC, 1992).

A second justification for focusing on tropical rain forests is that species living in tropical forests are more vulnerable to extinction. A species goes extinct when all its local populations have been extirpated. This is more likely when local populations or the species' geographic range, or both, are small. Small local populations are an inevitable outcome of high species diversity, which is a characteristic of tropical rain forests. Tropical species also tend to have smaller geographic ranges (Terborgh, 1974).

A third reason is the inability of many rain forest organisms to recolonize deforested land. A large proportion of the plants and animals are unable to cross open stretches to the next forest patch (e.g., Wells, 1988). Also, most birds and mammals of the rain forest have specialized feeding behavior and cannot survive human alteration of the forest (e.g., Cranbrook, 1988; Wells, 1988). For instance, strip-cutting has been proposed as a less damaging alternative to selective timber felling in tropical rain forests. However, a recent study (Gorchov, 1994) found that regeneration of animal-dispersed trees in the clear-cut strips was dramatically delayed, even though the strips were a mere 30–40 meters wide. Likewise, in order to generate and grow, many rain forest tree seedlings require specific sheltered microclimates provided by intact forests or by relatively small canopy gaps. In areas of poor soil fertility, large human-made clearings are reforested only slowly, at least in part because the removal of the original forest vegetation caused critical nutrients to escape from the ecosystem (Jordan, 1988).

These predisposing factors, then, conspire to make tropical rain forest species more prone to extinction than those of other habitats in the face of the traditional challenges accompanying economic development. The upshot is that tropical rain forests house a greater proportion of the world's threatened species of birds and mammals than any other type of habitat (WCMC, 1992).

High biodiversity and vulnerability are not the only justifications for focusing on tropical rain forests. Along with some boreal forests, tropical rain forests are the last intact major forest ecosystems on Earth. The subtropics were the first to sustain human agriculture, and their forests were leveled quickly (Ponting, 1991). The temperate zones followed suit, and virtually no pristine or old-growth forests are left in Europe, mainland East Asia, or North America east of the Rocky Mountains. In the tropics, the human population traditionally has been highest in the seasonally dry regions originally covered with a mosaic of deciduous and gallery forests. Very little of the original tropical dry forest now remains.

By contrast, the large-scale exploitation of tropical rain forests only began after World War II, and there is still a good amount of forest left, although estimates as to the extent of their disappearance and modification vary widely (Whitmore and Sayer, 1992). At least in certain regions, there is still an opportunity to save some of these forests in their wild and untampered state without having to go through costly attempts at habitat restoration. Hence, because of this unique set of problems and opportunities, tropical rain forests deserve our special attention. We should note, of course, that other biomes have totally different sets of species, so conserving all the rain forests in the world will not save, say, Mediterranean fynbos ecosystems or tropical dry forests. Thus, a global strategy should include protected areas representative of all the major ecosystems.

## PREVIEW OF THE CONTENTS

Successful solutions to the problems besetting protected rain forest areas require the effective integration of a number of different perspectives. An ecological perspective is needed to understand the extent of the biodiversity crisis and its implications for ecosystem functioning. A socioeconomic perspective is needed to understand why humans are so rapidly altering tropical forest landscapes. A management perspective is needed to determine how protected areas and buffer zones can be better managed to conserve biodiversity and remain financially viable. Finally, a policy and institutional perspective is needed to devise possible reforms at the local, national, and international level—reforms that would provide stronger incentives to improve the global network of protected areas in the tropics.

In chapter 2, Terborgh and van Schaik review the evidence for the presumed crisis of tropical biodiversity, and they refute some of the more commonly expressed objections to the idea that there is a mass extinction in the making. They also point out that protected areas are the cornerstone of any strategy of preserving biodiversity, given the biological and social problems surrounding sustainable use and the increased pressure on resources in virtually all areas subject to some form of use.

Chapters 3 and 4 review the world's protected rain forest areas. MacKinnon considers the ecological problems faced by protected area systems currently in place and concludes that coverage is adequate in some areas but far from adequate in others. She also identifies many design problems that need to be improved in the future if these protected areas are to continue to function as viable ecosystems that maintain their biodiversity in the long run. Conservation planning is for the long haul.

Van Schaik, Terborgh, and Dugelby then examine the social threats to the world's protected rain forest areas. Their unwelcome conclusion is that virtually all protected areas face one threat or another, often several, regardless of the affluence of the countries involved. After reviewing the root causes, they conclude that poverty is one factor, along with increased population pressure (mainly locally), but that large-scale conversion and exploitation by nonlocals is also important in places and is much less recognized.

Brandon (chapter 5) surveys the project-level impacts of the sustainable-use paradigm of protection. In particular, she examines a number of integrated conservation-development projects (ICDPs) to determine how successful they have been in meeting a variety of goals. Her conclusions are not encouraging: ICDPs have by and large failed to improve the maintenance of the biodiversity values they set out to preserve. However, she provides a number of recommendations for improving the design of such projects.

In chapter 6, Sanderson and Redford examine the expanding set of actors that have a stake in the debate about biodiversity conservation. Some of these actors have seized the biodiversity discourse as a means to achieve goals that have little to do with, and in fact may conflict with, biodiversity conservation. They argue that much of the debate centers around property rights: who owns biodiversity? They conclude that the scale at which biodiversity is defined has important implications for appropriate political structure and management choices.

Miranda and LaPalme (chapter 7) review the social norms that influence the possibilities for using protected areas to protect biodiversity in the tropics. They examine how different forest property regimes—private, communal, open access, and state ownership—can influence use of biological resources in positive and negative ways. They also examine how agricultural and forestry policies, as well as other policies, affect land-use decisions by people living in the vicinity of parks.

Kramer and Sharma (chapter 8) review the flow of use and nonuse benefits that people derive from tropical forests, and they observe that the sustainable use movement has given inadequate attention to the importance of the nonuse benefits that the world community derives from biodiversity conservation in the tropics. They characterize biodiversity protection as a global environmental good with benefits that go far beyond the national boundaries of countries with protected areas. Since the establishment of protected areas often imposes significant costs on local people, some transfer of financial resources is therefore warranted.

In chapter 9, Ferraro and Kramer consider the issue of compensation (monetary and nonmonetary) for local residents who lose access to forest resources due to the establishment of protected areas. They consider arguments for and against compensation based on considerations of legality, efficiency, equity, ethics, and conflict mitigation. They conclude by discussing the desirable aspects of an approach that would create incentives for biodiversity conservation while avoiding some of the potential shortcomings of many compensation schemes.

Finally, in chapter 10, van Schaik and Kramer put forth a set of recommendations on where we should go from here. Given the past loss of biodiversity in the tropics, the increasing pressure on protected areas expected in the future, and the unhappy state of many of the world's protected areas, what can be done to improve these key elements of an overall strategy to conserve the biodiversity of tropical rain forests?

## NOTES

Thanks to Barbara Dugelby and Marie Lynn Miranda for comments. We benefited from discussions with Nick Salafsky, Priya Shyamsundar, and John Terborgh.

This work was supported by a cooperative agreement between the U.S. Agency for International Development and the Duke University Center for Tropical Conservation. Carel P. van Schaik's fieldwork is supported by the Wildlife Conservation Society.

## REFERENCES

Batisse, M. 1982. The biosphere reserve: a tool for environmental conservation and management. *Environmental Conservation* 9 (2): 101–11.

Cernea, Michael, ed. 1991. *Putting people first: sociological variables in rural development*, 2d ed. New York: Oxford University Press.

Cranbrook, Earl of. 1988. Mammals: distribution and ecology. In *Malaysia*, Earl of Cranbrook (ed.), 146–66. Oxford: Pergamon Press.

Gorchov, D. L. 1994. Natural forest management of tropical rain forests: what will be the "nature" of the managed forest? In *Beyond preservation: restoring and inventing landscapes*, A. D. Baldwin Jr., J. De Luce, and C. Pletsch (eds.), 136–53. Minneapolis: University of Minnesota Press.

Hays, Samuel P. 1987. *Beauty, health and permanence: environmental politics in the United States, 1955–1985*. New York: Cambridge University Press.

IUCN, UNEP, and WWF [World Conservation Union, United Nations Environment Programme, and World Wildlife Fund]. 1980. *World conservation strategy: living resource conservation for sustainable development*. Gland, Switzerland: IUCN, UNEP, and WWF.

IUCN, UNEP, and WWF [World Conservation Union, United Nations Environment Programme, and World Wildlife Fund]. 1991. *Caring for the earth: a strategy for sustainable living*. Gland, Switzerland: IUCN, UNEP, and WWF.

Jordan, C. F. 1988. *Nutrient cycling in tropical forest ecosystems; principles and their application in management and conservation.* Chichester: Wiley.

Kramer, Randall A., Robert Healy, and Robert Mendelsohn. 1992. Forest valuation. In *Managing the world's forests*, Narendra Sharma (ed.), 237–67. Dubuque, Iowa: Kendall/Hunt Publishing.

MacKinnon, John, Kathy MacKinnon, Graham Child, and Jim Thorsell. 1986. *Managing protected areas in the tropics.* Gland, Switzerland: IUCN (World Conservation Union).

McNeely, J. A. 1988. *Economics and biological diversity: using economic incentives to conserve biological resources.* Gland, Switzerland: IUCN (World Conservation Union).

Office of Technology Assessment, U.S. Congress. 1987. *Technologies to maintain biological diversity.* Washington, D.C.: U.S. Government Printing Office.

Peres, C. 1994. Exploring solutions for the tropical biodiversity crisis. *Trends in Ecology and Evolution* 9:164–65.

Ponting, C. 1991. *A green history of the world: the environment and the collapse of great civilizations.* New York: Penguin.

Primack, Richard B. 1993. *Essentials of conservation biology.* Sunderland, Mass.: Sinauer.

Public Law 64-235. 1916. *U.S. statutes at large*, vol. 39: 535. Washington, D.C.: U.S. Government Printing Office.

Redford, Kent H., and Steven E. Sanderson. 1992. The brief, barren marriage of biodiversity and sustainability. *Bulletin of the Ecological Society of America* 73 (1): 36–39.

Robinson, J. G. 1993. The limits to caring: sustainable living and the loss of biodiversity. *Conservation Biology* 7 (1): 20–28.

Robinson, John G., and Kent H. Redford. 1991. *Neotropical wildlife use and conservation.* Chicago: University of Chicago Press.

Schulze, E. D., and H. A. Mooney. 1993. *Biodiversity and ecosystem function.* Ecological Studies vol. 99. Berlin: Springer.

Terborgh, J. 1974. Preservation of natural diversity: the problem of extinction-prone species. *Bioscience* 24:715–22.

USAID [U.S. Agency for International Development]. 1992. *Tropical forests and biological diversity: USAID report to Congress 1990–91.* Washington, D.C.: USAID.

WCMC [World Conservation Monitoring Centre]. 1992. *Global biodiversity: status of the earth's living resources.* London: Chapman and Hall.

WCS [Wildlife Conservation Society]. 1993. *1991–92 annual report.* New York: WCS.

Wells, D. 1988. Birds. In *Malaysia*, Earl of Cranbrook (ed.), 167–95. Oxford: Pergamon Press.

Wells, Michael, and Katrina Brandon, with Lee Hannah. 1992. *People and parks: linking protected area management with local communities.* Washington, D.C.: World Bank, World Wildlife Fund, and U.S. Agency for International Development.

West, Patrick C., and Steven R. Brechin. 1991. *Resident people and national parks, social dilemmas, and strategies in international conservation.* Tucson: University of Arizona Press.

Whitmore, T. C., and J. A. Sayer. 1992. Deforestation and species extinction in tropical forests. In *Tropical deforestation and species extinction*, T. C. Whitmore and J. A. Sayer (eds.), 1–14. London: Chapman and Hall.

Wilson, E. O. 1992. *The diversity of life*. Cambridge, Mass.: Harvard University Press.
WRI, IUCN, and UNEP [World Resources Institute, World Conservation Union, and United Nations Environment Programme]. 1992. *Global biodiversity strategy: guidelines for action to save, study, and use earth's biotic wealth sustainably and equitably*. Washington, D.C.: WRI, IUCN, and UNEP.
Wright, S. J. 1992. Seasonal drought, soil fertility and the species density of tropical forest plant communities. *Trends in Ecology and Evolution* 7 (8): 260–63.

# 2

# Minimizing Species Loss: The Imperative of Protection

*John Terborgh and Carel P. van Schaik*

The major rush of extinctions under way in tropical rain forests is caused by habitat loss, habitat fragmentation, and overexploitation of species useful or threatening to humans. As burgeoning human populations continue to claim an ever greater share of the world's renewable resources, intensification of land use will make it increasingly difficult to maintain biodiversity outside of strictly protected nature preserves. Further, because many species can maintain themselves only in large expanses of unaltered or lightly disturbed habitat, fully protected areas should remain the cornerstone of any conservation strategy that aims at minimizing the loss of tropical biodiversity.

We have identified five principal forces of extinction in a human-dominated world: deforestation, habitat fragmentation, overkill, secondary extinction, and introduced species. In this chapter, we briefly review the manner in which each of these processes contributes to extinction, both as isolated forces and in synergism with other forces.

Our conclusion is that all five forces of extinction can be minimized by retaining intact natural habitat in the largest possible blocks. As the human population continues to expand, it is inevitable that most land outside strictly protected nature preserves will be subject to increasingly intensive use, resulting in decreased biodiversity. We therefore argue that maintaining biodiversity can best be achieved through development planning at the largest practical spatial scales. Parks should be as large as possible, designed to benefit from passive protection (inaccessibility), and rigorously protected. Only through major strengthening of institutions responsible for park protection can we expect to see tropical biodiversity survive the coming century.

## THE PRESSURES OF POPULATION GROWTH

The earth is experiencing an extinction crisis because the human population is increasing rapidly and laying claim to an ever larger share of land and resources. Simultaneously, nearly all individual humans fervently desire to increase their level of material well-being. The resulting double impetus for rapid economic expansion generates exponentially increasing demands for most renewable and nonrenewable resources. The world is consequently experiencing a wave of nonsustainable use of most basic, life-supporting resources, including soil, groundwater, forests, grasslands, and fisheries.

The earth's population today is 5.5 billion, and it is increasing by more than 90 million per year. By even conservative projections, the total is expected to surpass 10 billion before 2050 (Urban and Nightingale, 1993). We must therefore anticipate a near-term future in which demands on virtually all exploitable natural resources will greatly exceed current levels. Yet, in many cases, current levels of exploitation are clearly unsustainable (Postel, 1994). The blatant discord between the fixed capacity of natural systems to supply resources and the escalating demands of the human population seems likely to climax in the coming half-century.

The unavoidable transition from an exponentially growing economy, fueled by profligate consumption of resource capital, to a truly sustainable human lifestyle will not be accomplished without severe global stress. This period is likely to be a time of unprecedented social disorder, accompanied by the breakdown of human institutions in overcrowded parts of the world (Kaplan, 1994). The turmoil we are witnessing today in such countries as Somalia, Rwanda, Angola, Cambodia, and Haiti may only be a foretaste of the future. As poverty and resource depletion continue to widen, countries all over the globe are certain to experience similar convulsions. Whether any significant remnants of tropical nature will survive the coming period of upheaval is an open question.

It is against this backdrop that we discuss the extent of the current erosion of biodiversity in the tropical rain forest biome, examine its causes, and develop realistic options for the future.

## WHAT IS BIODIVERSITY?

Biodiversity, quite simply, is the total number of species occupying a region, continent, or the entire planet. The emphasis is on species because species are the units of evolution and because species (not populations, varieties, ecosystems, or other subdivisions of nature) go extinct. It is the finality of extinction that underscores the importance of species. The centrality of species in the definition of biodiversity is implicitly recognized in the Endangered Species Act, currently the leading instrument of biodiversity protection in the United States.

Some opponents of the Endangered Species Act have argued that extinction is a normal part of the evolutionary process, and that the imposition of restraint for the sake of preventing something that will happen anyway is nonsensical. This argument is false, however, because these opponents invariably fail to mention that in evolution, extinction is normally balanced by speciation, the gradual origination of new species, with the result that diversity is maintained. For extinction rates to rise manyfold with no concurrent increase in speciation is decidedly unnatural. Since speciation cannot be accelerated, the effort to conserve the earth's biodiversity for future generations must therefore focus on strategies for minimizing extinction.

The current rapid rate of extinction is especially worrisome in the world's tropical rain forests. Covering less than 7 percent of the earth's land surface, these forests harbor more than 50 percent of all known species, including those that occupy both fresh and salt water. Why biodiversity is so disproportionately concentrated in a single ecological formation remains an incompletely answered scientific question. Nevertheless, the very fact that rain forests are the habitat of more than half of all species compels us to grant them the topmost priority in any global strategy to preserve biodiversity.

## ESTIMATES OF SPECIES LOSS

The claims of an impending extinction spasm that have appeared prominently in recent years—notably, in the fundraising literature of several conservation organizations—have been criticized by skeptics as baseless doomsaying. Critics have pointed to the incongruous juxtaposition of the very low number of recorded extinctions and the seemingly astronomical estimates of current species loss. The disparity in the numbers does, indeed, require explanation.

From approximately 1600 A.D. to the present, only about 600 species of plants and 500 animal species have officially been declared extinct (WCMC, 1992; Smith et al., 1993). The numbers are incongruously low because of the very formidable practical difficulty of proving extinction. How long does one wait after the last credible sighting of a species before extinction is acknowledged? Even in the United States, where the availability of information is excellent compared to most tropical regions, ambiguity can linger for decades. Persistent reports of ivory-billed woodpeckers continue to leak out of Florida, Louisiana, and Texas, although the last unequivocal photographs of the species were taken in the 1950s. Is the ivory-billed woodpecker extinct? It is a matter of opinion, because the absence of something cannot be proven. Couple this fundamental ambiguity with the near vacuum of information that exists for many tropical countries, and the problem of estimating extinctions becomes apparent. Poor baseline information may engender a false sense of optimism about the extinction problem. The species most likely to be on record as present—that is, the common ones—are also the most likely to survive.

It is only where biologists have systematically repeated surveys that losses are apparent. In West Malaysia, for instance, fewer than half of the freshwater fishes historically known to occur in the region were found during a four-year survey in the early 1980s (Wilson, 1992). In the state of Perak in West Malaysia, many of the endemic tree species collected in the 19th century have never been found again despite regular surveying by foresters. In the meantime, most of the lowland rain forest of Perak has been converted into rubber plantations (Whitmore, 1990).

Similarly, Gentry (1986) discovered 38 narrowly endemic plant species in the Centinela ridge in the cloud forest zone of western Ecuador, and suspected that further study of the collections would reveal another 60 or so endemics. However, a resurvey provided photographic documentation that all the forest on the ridge and in the surrounding landscape had been razed. Subsequent collecting in nearby areas has failed to reveal the presence of the formerly endemic species.

Another way to appreciate the scale of the extinction problem is to consider organisms that have not yet gone extinct. It appears that the better a taxonomic group has been studied, the higher the proportion of species that are considered threatened. For example, the taxonomy and distribution of palms are relatively well understood, and it has been estimated that 33 percent of palm species are threatened. Much less is known about the taxonomy of insects or crustaceans, and the percentages of these species threatened are estimated to be 0.07 and 3, respectively (Smith et al., 1993). Detailed local surveys suggest that even these percentages are still very conservative. We can surmise that many species must have disappeared before they were ever collected, discovered, and described. With countless biological species still awaiting to be discovered, we simply started too late to be able to document species extinction with any precision.

Scientific ignorance of the current status of a majority of the earth's species has led to other sources of confusion. Widely varying estimates of species loss (Reid, 1992) have been construed as evidence of wide disagreement on the extent of the biodiversity crisis. Instead, inconsistency reflects the sad fact that these estimates cannot be more than very crude guesses, as they must be based on a host of unverifiable assumptions. These assumptions include the total number of species initially present in a habitat; the slope of the species-area curve, which indicates the relationship between the number of species and the area of land; the extent of deforestation or habitat degradation; the degree to which the new habitats are hospitable to forest species; the extent of forest fragmentation; and the time scale of secondary extinctions. The exact number of species lost this century will never be known, but we do know that the number is certainly large.

The main point we wish to emphasize is that a lack of knowledge of the extent of the biodiversity crisis can never be used as an excuse to do nothing. All indications are that a major loss of species is already in progress and that the worst is yet to come.

Our goal for the future is to retain as large a fraction of Earth's biodiversity as possible. Although we realize that many more species will inevitably be lost in the coming decades, preserving biodiversity to any significant degree requires that conservation strategies be designed explicitly to minimize extinction. It is therefore critically important to understand the extinction process and the forces that promote it.

## LEADING CAUSES OF EXTINCTION

The causes of extinction are well known to biologists. The proximate cause is nearly always extreme rarity, coupled with reproductive insufficiency. Although a substantial research effort has been devoted to understanding the special biology of small populations, it generally will not be practical to apply the knowledge to tropical biodiversity conservation. Once a species has suffered such a drastic decline that it is subject to small population effects, such as inbreeding, it is probably beyond recall (Schonewald-Cox et al., 1983; Soulé, 1987). Rescuing a species from the brink of extinction often requires a concerted and expensive recovery program, such as those the United States has mounted to save the whooping crane and the California condor. With the exceptions of rhinoceroses, okapis, tigers, and a few other glamour species, few of the myriad inhabitants of tropical forests will receive such special attention. It is therefore more appropriate and useful to consider ultimate, rather than proximate, causes of extinction.

The ultimate causes of extinction in our human-driven contemporary world have been likened to the four horses of the apocalypse: habitat loss and fragmentation, overkill, secondary extinctions, and introduction of exotic species (Diamond, 1984). Here we treat the effects of habitat loss and fragmentation separately, thereby acknowledging the threats of five forces of extinction. Climate change may become a sixth force in the next century, but it is not now recognized as a major contributing factor, except possibly in the case of declining amphibians (Blaustein and Wake, 1990). We shall briefly consider each of the five forces in turn.

### Habitat Loss

Nearly all habitats everywhere are being degraded or converted wholesale to human-made landscapes. Tropical rain forests are predicted to endure outside of protected areas for only some 35 or 40 more years, as illustrated in Figure 2-1 (Terborgh, 1992a). Pristine, old-growth lowland tropical forests of West Africa, the Greater Antilles, Madagascar, India, the Philippines, and Atlantic Brazil—all rich in endemics—have already been reduced to less than 10 percent of their original areas (NRC, 1980).

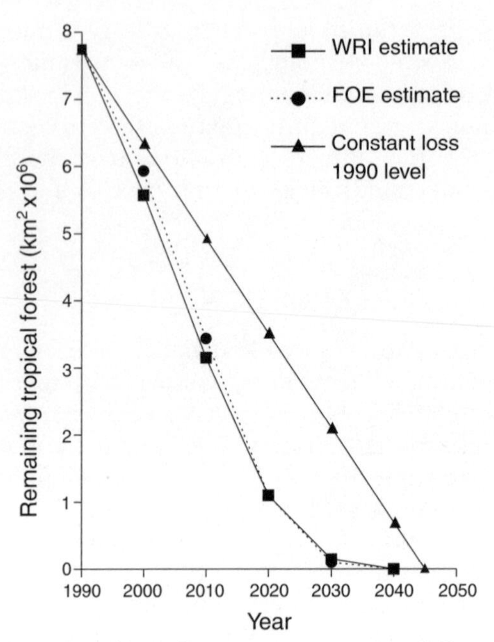


***Figure 2-1*** Projections of rain forest cover assuming different deforestation scenarios. *Source:* Terborgh, 1992c.

The progressive loss of tropical forests in Costa Rica, for example, is illustrated in Figure 2-2. Although Costa Rica has done more than any other tropical country to conserve its natural resources, its deforestation rate is one of the highest in the Americas, and little forest is predicted to remain outside of protected areas by the end of the decade (Terborgh, 1992c). Despite the increased worldwide attention to the loss of tropical forests, there is no indication that the rate of loss or degradation is slowing down. In fact, recent analyses indicate that the rate is increasing (Aldhous, 1993). In one study, for example, tropical deforestation was found to be going up in 14 of the 16 countries studied (Whitmore and Sayer, 1992).

Tropical rain forests are being converted for a variety of immediate reasons, but the fundamental cause is always the same: the forest itself or the land under it offers natural resources that can be profitably exploited for subsistence or commercial gain. In the poorer parts of the tropics, large-scale conversion is bound to continue as long as hundreds of millions of unemployed people have no other way to survive than to engage in swidden (formerly called slash-and-burn) agriculture. Sadly, alternative employment for a significant fraction of the world's population is not in prospect.

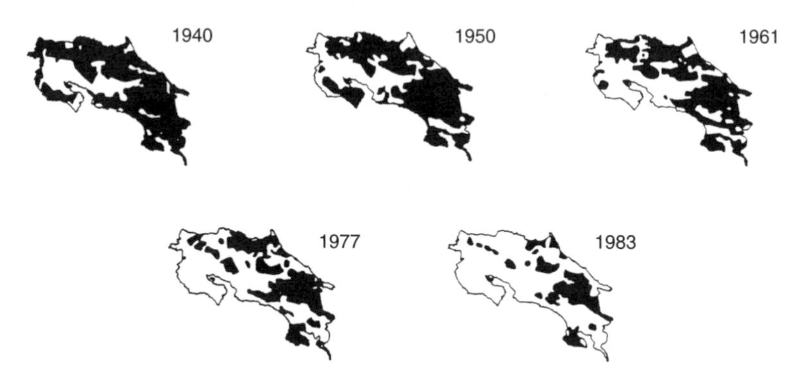


*Figure 2-2* The progressive deforestation of Costa Rica, as derived from historical and satellite data. *Source:* Terborgh, 1992c.

## Fragmentation

Fragmentation is a concomitant of habitat loss almost everywhere. Fragmentation restricts plants and animals to "islands" of habitat, often too small to sustain viable breeding populations (Terborgh, 1992b). A much-quoted rule of thumb holds that a 90 percent reduction in the area of a given habitat will result in the immediate loss of about half of the species contained in the habitat (e.g., Wilson, 1992). The rule of thumb is based on the species-area curve produced by surveying increasingly larger portions of a continuous natural environment.

However, by simplistically taking a static view of an intrinsically dynamical process, the rule of thumb results in a gross understatement of eventual species loss. Once fragments have been isolated, species loss continues—often rapidly—through the process known as secondary extinction (discussed below). The final equilibrium state, which may not be attained for decades or centuries, is one of severe biological impoverishment (see Figure 2-3). Depending on the size of the remaining fragment, 90 percent or more of the initial species complement can disappear (Diamond, 1984).

A few authors have disparaged such predictions, claiming that the island analogy is invalid (Simberloff and Abele, 1976). However, accumulating evidence has reaffirmed that the analogy is indeed germane: forest fragments in a human-made landscape display dynamics of species extinction and immigration very similar to those of true islands (Terborgh and Winter, 1980; Wilcox, 1980; Lovejoy et al., 1986). For many obligate forest species, a "sea" of clear-cut land not only is uninhabitable terrain but also poses a formidable barrier to dispersal and hence to the reestablishment of locally extinct populations. As the natural biodiversity of fragments gradually dwindles, species typical of adjacent altered habitats invade. Eventually, fragments contain only those

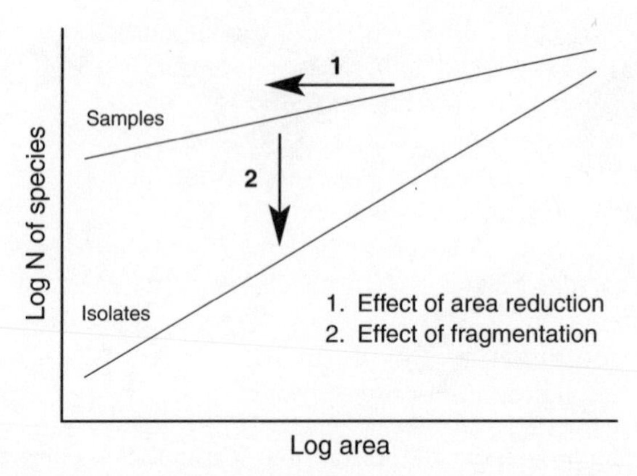


*Figure 2-3* The effects of habitat loss and fragmentation compared. The upper curve represents the number of species found in "samples," areas that are part of a larger matrix of unmodified habitats. The lower curve represents the species richness found in "isolates," areas not connected to other patches with unmodified habitat and surrounded by inhospitable habitat.

species that can survive in—and hence move through—the surrounding habitats (Janzen, 1986).

Studies of forest fragments conducted in various parts of the world reveal some common denominators in the pattern of species loss. Large birds and quadrupedal mammals, especially those favored by hunters, are often among the first to vanish, in company with top predators, nomadic frugivores, and naturally rare species (Terborgh and Winter, 1980; Diamond, 1984). Once locally extirpated, quadrupedal mammals are slow to recolonize because they disperse across inhospitable habitat more slowly than their flying counterparts, birds and bats. Not surprisingly, then, the highest proportion of endangered species is found among mammals (WCMC, 1992).

## Overkill

Overkill is defined as the decimation or extinction of wild populations of plants and animals through overharvest. Overkill was responsible for the demise of the mammoths and mastodons of North America, the elephant birds and giant lemurs of Madagascar, the giant kangaroos of Australia, the moas of New Zealand, the ground sloths and glyptodonts of North and South America, and many of the native birds of Hawaii and other Pacific Islands (Olson and James, 1982; Diamond, 1984; Martin and Klein, 1984; Wilson, 1992). These examples of overkill can be attributed to what we condescendingly refer to as "primitive" humans. In postcolonial North America, humans of European origin eliminated the passenger pigeon, sea mink, Steller's sea cow, Labrador duck, heath hen, and Carolina parakeet and came precariously close to imposing the same fate on the American bison, Thule elk, red wolf, and whooping crane (Ehrlich and Ehrlich, 1981).

Overkill typically precedes habitat destruction and has already emptied much of the vast Amazon forest of any animal larger than a squirrel (Redford, 1992). The same can be said of much of what remains of the West African forest. Market hunting and subsistence hunting are freely allowed in both regions, but of the two, market hunting is by far the more pernicious. Overkill, clearly, is not a thing of the past.

Nor is overkill perpetrated only by the economically destitute trying to make ends meet. In contemporary times, rhinoceroses are being reduced to the brink of extinction for their horns; giant pandas and giant otters for their skins; black caimans for their hides; tigers for their skins and just about every bone in their bodies; and manatees, giant river turtles, and many other species for their meat. Similarly, several tree species with prized timber (e.g., mahogany in the neotropics, rosewood in Madagascar, ironwood in Borneo and Sumatra, ebony in Sulawesi) have been exploited so systematically that they are almost extinct in the wild.

International trade in wildlife products is one of the most lucrative of all illegal activities. Although the Convention on International Trade in Endangered Species (CITES) treaty has helped to slow international commerce in endangered species, it needs strengthening—the profits are so enormous, and the penalties meted out to violators so mild, that there is now little deterrence for the professional criminal organizations that have taken over this trade.

For the most part, however, international treaties do little to regulate domestic commerce in endangered species. In the absence of more encompassing protective legislation, overkill can be expected to continue as a major driving force in extinction.

## Secondary Extinctions

Secondary extinctions are a ubiquitous and especially pernicious concomitant of habitat fragmentation—pernicious because fragments undergoing secondary extinctions may appear perfectly intact and healthy to an untrained observer (Soulé et al., 1988). Secondary extinctions are defined as losses of animal or plant species resulting directly or indirectly from the abnormal abundance or absence of other species. Such losses are expected where there are immediate and obvious dependencies—for instance, of plant-eating insects limited to only one species of tree. When the tree goes extinct, the insect follows (Gilbert, 1980). Other dependencies are less obvious but no less critical. A notorious case is that of a tree found only on the island of Mauritius in the Indian Ocean. The fruits and seeds are large and could only have been dispersed by the dodo, the largest animal on the island prior to the arrival of European sailors in the 17th century. Only a small number of ancient and senescent trees now remain alive. In the absence of dodos, the trees have no possibility of replacing themselves except through the intervention of humans (Temple, 1977). Other such examples are gradually coming to light, involving, for instance, gorillas (Tutin et al., 1991) and forest elephants (Martin, 1991).

The topic of secondary extinctions is currently under active investigation, and as more scientific information becomes available, the more ubiquitous the driving forces are proving to be. Even the eastern United States is in the midst of a cascade of extirpations and potential extinctions, though the public is generally unaware of the severity of the situation. The systematic extermination of top predators (wolves and mountain lions) a century or two ago has greatly reduced the natural mortality of several prey species, including, among others, the white-tailed deer. These deer, as suburbanites all over the eastern United States can attest, have become abnormally abundant, and in some areas live at densities several to many times higher than in natural communities containing top predators. Excessive browsing by deer is suppressing or altering the pattern of forest regeneration over large areas and severely threatening a number of endangered plant species (Alverson et al., 1988; Miller et al., 1992). In time, a superabundance of deer will significantly alter the composition of eastern forests, and as the tree composition changes, so will the associated birds, insects, herbaceous plants, and other organisms.

Ominous distortions of predator-prey and plant-herbivore interactions are already a fact in superficially pristine primary forests all over the tropics, as natural ecosystems are perturbed by systematic overhunting of large birds and mammals. Among the most heavily affected functional groups are top predators, large herbivores, seed predators, and seed dispersers. Although little studied to date, the fundamental alterations of ecosystem function that accompany the loss of species that mediate such important biological processes can be expected to result in cascades of secondary extinctions, as the absence of key species causes interaction webs to collapse (Terborgh, 1988; Dirzo, 1990).

The key role played by top predators, such as tigers and jaguars, in regulating prey densities in tropical forests has only recently been scientifically recognized. Carnivores maintain the populations of functionally important prey species—such as herbivores, seed predators, and seed dispersers—at densities well below the levels that could be sustained by the available food supply. Therefore, thanks to the action of carnivores, some seeds escape the depredations of seed predators, such as pigs and rodents, and some seedlings escape the notice of herbivores, such as ungulates. The forest is thus able to regenerate in its full diversity, in accordance with a natural balance of predator and prey that has prevailed throughout evolutionary time (Terborgh, 1992c). But where the carnivores are persecuted—and that is almost everywhere—then, as we have seen in the case of the white-tailed deer, the natural system of checks and balances breaks down, and the whole system goes into decline.

It thus becomes a matter of the utmost importance to retain top carnivores in areas where biodiversity is being conserved (Terborgh, 1992b). Unfortunately, top carnivores require extravagant amounts of living space. A female jaguar, for example, needs 30–50 square kilometers to support herself and her dependent offspring. A population of jaguars of minimum viable size, perhaps 500 individuals, therefore requires on the order of 10,000 square kilometers (1 million hectares). Only a few rain forest parks are as large as this. In smaller areas, there is a serious risk of losing the top carnivores and thereby triggering a cascade of secondary extinctions.

## Introduced Species

Global free trade fostered by the General Agreement on Tariffs and Trade (GATT) treaty can be expected to stimulate a massive increase in intercontinental commerce. Increased intercontinental movements of goods and people will inevitably result in both intentional and inadvertent introductions of exotic species to all continents and inhabited islands, as well as to all the world's oceans and seas (Carlton and Geller, 1993; Hedgpeth, 1993). Among the species to be introduced will be pests, pathogens, predators, herbivores, and invasive weeds, both plant and animal (although the lexicon inconveniently lacks a term for weedy animals, such as the European house sparrow and common carp).

If the past is any guide, many future introductions will have undesirable consequences, including extinctions. Examples abound, among them the inadvertent introduction of diseases that have decimated the American elm and American chestnut, two of the most important tree species of the eastern United States. Now, the gypsy moth is ravaging these same forests, with a consequent loss of millions of dollars worth of standing timber.

Introduced species have been particularly destructive in freshwater aquatic systems and on oceanic islands (Zaret and Paine, 1973; Zaret, 1980; Lowe-McConnell, 1987; Atkinson, 1989). For example, the brown tree snake, accidentally introduced to Guam during World War II, has decimated the island's native birds (Savidge, 1987). At least two of these, the Marianas kingfisher and Guam rail, now exist only in captivity. Mosquitos, carried to the Hawaiian islands during the 18th century in ships' water supplies, later transmitted avian pox and avian malaria from introduced birds to native birds, causing the extinction of a host of species and the severe endangerment of several more (Warner, 1968). Introduced goats and pigs have denuded many islands of their natural vegetation, including several of the Galapagos Islands (Atkinson, 1989).

Outside of the polar regions, there is hardly an island that has not suffered from the ravages of introduced species. The obvious remedy is to prohibit any further introductions of alien species, but rigorous implementation is simply not a realistic possibility, given the realities of global trade. Additional unplanned and unwanted introductions will continue to impose a high cost in extinctions and economic losses for the foreseeable future. Nevertheless, this reality should not deter us from exerting every possible effort to minimize the losses.

## Interactions Among the Forces

We have now described the five major forces that are propelling the current extinction crisis. It is important to note that the forces do not operate independently of one another. As mentioned above, overkill typically precedes or accompanies habitat loss, as local people scour the forests within walking distance of their villages in search of game or marketable animal products. Later, as the agricultural frontier expands into the forest, the most suitable sites are cleared first, resulting in a fragmented landscape (Figure 2-4). As more settlers arrive, and clearing continues, the forest melts away and eventually only small, biologically inviable fragments remain, typically in deep ravines and on ridge tops. The fragments then experience cascades of secondary extinctions, frequently exacerbated by the invasion of nonforest and exotic species from adjoining disturbed areas. At this point, the landscape retains little value from the standpoint of natural biodiversity. There is thus a perverse synergism between the five driving forces that can exceed the sum of the independent effects.

*Figure 2-4* Overexploited tropical rain forest fragment (Karo Highlands, Sumatra, Indonesia). Photo by Chris Hildreth.

## COMPREHENSIVE LAND-USE PLANNING AND PROTECTED AREAS

Although the five forces of extinction operate through biologically distinct mechanisms, amelioration of all of them can effectively be accomplished by a single means: planning and implementing development at the largest possible spatial scales. The key criteria are to concentrate development in the most appropriate areas—those offering good soil, level ground, adequate water supply, and so on—and to forego the building of roads wherever possible outside designated development zones. Put in these terms, the often-presumed conflict between conservation and development need not be great.

By retaining natural habitat in large roadless blocks, all five forces of extinction are addressed at once. Overkill is minimized in areas too large to traverse on foot. Unplanned and illegal habitat losses are most effectively limited in large tracts that minimize the exposed perimeter in relation to the included area. Fragmentation is avoided through the retention of intact blocks of habitat. And by preventing fragmentation, secondary extinctions and the invasion of exotic species are held to the lowest practical levels.

In the past, large-scale rural development projects, such as the Polonoroeste Project in Brazil, Highways of Penetration (into the Amazon) in Peru, and Transmigration in Indonesia, have been motivated by political concerns and implemented with little attention to scientific criteria or sustainability (Goodland and Irwin, 1975; Fearnside, 1986). A willy-nilly attitude toward the environment has led to some notorious environmental travesties and, in some of the most embarrassing cases, withdrawal of financial support by the World Bank and other lending agencies.

In a world of shrinking per-capita renewable resources, neither nature nor humanity can afford the luxury of further trial-and-error approaches to rural development. Nor is there any excuse for it. The science needed to achieve sustainability is available. A step in the right direction is the World Bank's recently adopted program to produce "National Environmental Action Plans"—country-by-country planning documents that attempt to anticipate future environmental pressures and needs and the institutional structures required to meet them.

Scientific land-use planning must be an integral part of the process of achieving sustainable development. In the context of biodiversity, the key criteria are the intensity and scale of land use, rather than the particular economic activity that may be practiced.

### The Land-Use Cascade

There is no single "best" use for a given tract of land. How land is used depends on a complex of economic and social factors. Historically, land use in many parts of the world has been highly dynamic, shifting, say, from rural to urban, or from forest to cropland, in response to economic and demographic pressures. In the rapidly developing contemporary world, the dynamics of land use shows a strong directionality, as illustrated by Figure 2-5.

The directionality is apparent if we examine land-use trends in relation to a simple, four-level categorization based on intensity of use and degree of alteration of the original habitat:

- *Wildlands*. These areas have unaltered or lightly altered natural vegetation harboring intact or only mildly perturbed animal communities. It stands to reason that in today's overcrowded world, wildlands remain only in regions possessing low to negligible potential for intensive development. Parks and nature reserves are most often sited in such "leftover" remnants of the presettlement landscape. Globally, wildlands are rapidly dwindling, but remaining areas, if given adequate protection, can provide important environmental services, such as atmospheric mitigation and watershed protection, in addition to conserving biodiversity.
- *Extensively used land*. Areas included under this heading are swidden agriculture, selectively harvested multispecies production forests, natural grasslands subject to livestock grazing, extractive reserves, public hunting grounds, and lightly used park buffer zones and watershed protection areas. Secondary forests and other vegetation recovering from disturbance would also fall into this category. Land subject to extensive use is often intrinsically of low quality and noncompetitive for more intensive uses (such lands, e.g., may have hilly or rocky terrain, poor soils, or arid or semiarid climate). New lands are requisitioned for extensive use mostly from remaining wildlands.

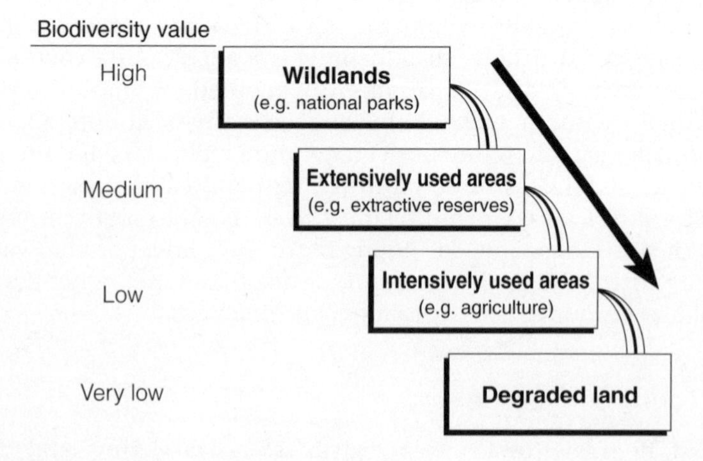


*Figure 2-5* The land-use cascade.

- *Intensively used land.* These include urban and industrial areas, croplands, pastures of nonnative grasses, orchards, and tree plantations. Land in this category is usually derived from areas under extensive use. In developed nations, economic competition results in the allocation of rural land among the agriculture, livestock, and forestry sectors in accordance with the principle of maximization of return on investment (Healy, 1985). In the poorest countries, however, millions of people live partly or entirely outside the market economy. The exigencies of subsistence compel them to exploit whatever land is available, regardless of its inherent suitability. Abused, exhausted, and abandoned land is consequently accumulating in nearly every country, as poor land-management practices deplete the soil and expose it to erosion.
- *Degraded land.* Degraded land is derived from lands previously under intensive or extensive exploitation. Invariably, degradation results from overexploitation driven by increasing human populations and aspirations. Misuse of land is dangerously antithetical to the goal of sustainable development. In fact, one criterion of sustainable development would be a stable land-use spectrum, just as a nongrowing population is characterized by a stable age distribution.

Land degradation is a global problem of alarming proportions. Worldwide since 1950, an area equivalent to the combined national territories of China and India has been abandoned (Postel, 1994). Land degradation deprives people of potential livelihoods, forecloses options for the future, and, in severe cases, can impel mass migrations, often across international boundaries, to urban centers or still unexploited wildlands (Terborgh, 1992c). In numerous countries, including Haiti, Madagascar, and Ethiopia, large expanses of once productive land have eroded down to bedrock, precluding any reasonable possibility of rehabilitation. In the seasonal tropics, particularly in southeast Asia, swidden agriculture has encouraged the spread of unpalatable alang-alang grass. Once established, such secondary grassland is perpetuated by fire and is highly refractive to reforestation efforts.

Nevertheless, given enough time, all transitions between stages of land use are potentially reversible. But given current global trends, the transitions are consistently downward (as depicted in Figure 2-5 by the arrow). Overall, we are witnessing a massive conversion of virgin habitat to degraded and abandoned land, as resource capital is consumed in an often brief wave of intensive exploitation.

Biodiversity suffers because it becomes increasingly reduced down the cascade. Habitats that are used as intensively as a Dutch cow pasture, an Iowa cornfield, or a Georgia pine plantation contain no more than a tiny fraction of their original biodiversity. Moreover, many of the species currently found in pastures, cornfields, and pine plantations are intentionally introduced exotics or escaped weeds. The vegetation of degraded land is often composed of a small number of weedy species, many of them introduced or cosmopolitan.

Most of the world's tropical forest occurs in countries experiencing rapid population growth. The expanding needs and aspirations of growing populations can only be met through the intensification of land use. There is no alternative. Both economic incentives and population growth drive the land-use cascade downward toward the accumulation of degraded land and biological impoverishment.

Intensification of land use will occur at the expense of remaining pristine forests and forests currently under extensive use. Nearly all of the land comprising the latter two categories is marginal for cultivation, so the amount of degraded land will continue to accumulate (Brown, 1994). The total amounts of land in the top compartments of Figure 2-5 will therefore decrease, probably sharply, while the amounts in the bottom two compartments will increase by the same amount. The decrease in the category of extensively used land will be caused by the conversion of seminatural multispecies forests into tree plantations, managed pastures, or croplands.

Retarding, and eventually reversing, the land-use cascade, along with population stabilization, should become the highest priorities of the multilateral and unilateral donor agencies. Although costly, and often technically difficult, the rehabilitation of abandoned land is currently regarded by most governments as only an option. In the future, it will become a necessity.

## IMPLICATIONS FOR CONSERVATION STRATEGIES

To maintain biodiversity, large areas must remain forested, ideally in contiguous tracts. One strategy for conserving forests would be simply to urge tropical countries to stop converting their rain forests and leave them alone. While this would certainly be best for biodiversity, it is not a realistic option, either socially or economically.

The population of the developing world is predicted to double in the next 30 years (Urban and Nightingale, 1993). It will consequently be impossible to halt the land-use cascade in the short run. A continuing trend toward the intensification of land use therefore seems inevitable. This compelling reality has led many conservationists to advocate the extensive use of tropical forests—for example, for game, fish, fruits, nuts, oils, latex, herbal remedies, local construction materials, and nonconsumptive uses such as ecotourism.

Advocates of extensive use correctly note that much tropical biodiversity has survived to date in lightly used forests, in which subsistence-level activities have been the norm. Lightly used forests are almost as effective for biodiversity preservation as was the original primary forest, especially where areas of occupation are still embedded in an old-growth matrix. The presumption that light use can continue to serve the purpose of biodiversity conservation, however, presupposes a static world. Extensive use of forests without species loss is usually achieved only under a limited set of circumstances, and these circumstances are becoming increasingly rare: low human density, subsistence rather than commercial use, and rudimentary technology (Arvard, 1993). Mounting social, economic, demographic, and political pressures combine to compel overexploitation and the use of an increasingly broad array of species.

A fatal flaw in the extensive-use paradigm is that it admits of no discernable limits: the watershed between sustainable and unsustainable use is undefined and can be ascertained only through painstaking investigation. Rarely are such investigations undertaken (Salafsky et al., 1993). Given mounting pressures for the intensified use of natural resources worldwide, it seems unlikely that most extensive-use systems will remain within the bounds of sustainability.

Enthusiasm for the extensive use of tropical forests has, in part, arisen from speculations that the harvest of nontimber products from intact forests can yield greater economic benefits than those to be derived from converting the land to nonforest uses (Myers, 1984; Peters et al., 1989). Where such speculations have been subject to critical scrutiny, they have been found to rest on shaky economic assumptions (Schwartzman, 1989; Terborgh, 1992c; Salafsky et al., 1993). Moreover, advocates of extensive use brush aside the overriding fact that primary tropical forests contain significant stocks of resource capital in the form of marketable timber and virgin soil. The lure of unexpended resource capital will continue to be a nearly irresistible temptation for all the major players in environmental dynamics: governments, big business, and local residents. We are therefore skeptical that the extensive use of multispecies tropical forests is anything other than a transitory phase in the land-use cascade.

For this compelling reason, we assume that whatever biodiversity remains in the tropical forest biome one or two generations from now will be contained only in effectively protected wildlands. Protected areas will therefore be the foundation of biodiversity protection in the future.

Strong grassroots support for nature protection clearly exists in the industrialized countries, and within these countries, areas set aside as parks benefit from the attention of citizen watchdog groups and a solid commitment to rigorous enforcement of implementing legislation. In many developing countries, however, protected areas tend to be unpopular, especially with politically powerful interest groups intent on the commercial exploitation of natural resources (Durning, 1993). Parks are widely viewed by impoverished residents, most of whom will never visit one, as creations of the central government that mainly benefit rich foreigners. Residents of nearby rural communities resent the establishment of a park as an arbitrary closing of the frontier, a foreclosure of future opportunities for their children. Parks are therefore under mounting pressure in the developing countries of the tropics.

The international conservation community has responded to these pressures by inaugurating new programs designed to seek accommodation between the needs of humans and the needs of nature. While the sentiments behind such efforts are understandable, such an accommodation admits of no clearly drawn line in the sand. Where do the needs of humans end and the needs of nature begin? In the absence of an explicit and unarguable answer to this question, social pressures will invariably encourage greater exploitation of natural resources.

Yet the message remains clear: it is an observable fact that biodiversity is maintained when nature remains intact over large areas. Large areas of unprotected natural habitat will, however, soon be a thing of the past. Pressures for increasingly intensive land use, fed by burgeoning human populations and unsustainable land-use practices, will see to this. Strictly protected areas must therefore serve for the foreseeable future as the last bastions of nature. Rigorous protection of parks should thus become the highest priority of efforts to conserve nature.

The division of future landscapes between areas reserved for preserving nature and areas allocated to human use will largely be determined by political processes. Already, the political good will that has led to the creation of hundreds of parks worldwide has largely been dissipated. The pace of establishment of new parks has slowed markedly in the tropics as a whole, and to near zero in parts of tropical Asia and Africa. Nominally, about 5 percent of the world's rain forest has been preserved. The diminishing rate at which new areas are being approved by legislatures suggests that the final total will be in the range of 6–7 percent.

Many existing rain forest parks are in precarious straits. All too often, impoverished governments are unable or unwilling to implement the enabling legislative acts, and the parks quickly turn into "paper parks," unprotected by any formal apparatus (see chapter 4). Where the responsible institutions are unable to function effectively, due to limitations of budget or personnel, the world community has often responded with assistance on an ad hoc basis. For the long run, however, the reactive approach is almost certain to fail. A major and sustained effort aimed at instilling commitment to protecting nature preserves will be indispensable to achieving the objective of minimizing future losses of tropical biodiversity.

At the same time, governments that seriously embrace the concept of sustainable development must begin a process of comprehensive science-based land-use planning and back it up with appropriate legislation and enforcement. Not to do so is tantamount to abandoning the future. Sustainable development absolutely requires, among other conditions, the halting and eventual reversal of the land-use cascade. Although technically challenging, soil and vegetation rehabilitation can restore some level of utility to much of the vast acreage of abandoned land around the world. Still, the best solution will be to protect rain forests before they are damaged or destroyed.

## NOTES

This work was supported by a cooperative agreement between the U.S. Agency for International Development and the Duke University Center for Tropical Conservation. Carel P. van Schaik's fieldwork is supported by the Wildlife Conservation Society.

## REFERENCES

Aldhous, P. 1993. Tropical deforestation: not just a problem in Amazonia. *Science* 259:1390.

Alverson, W. S., D. M. Waller, and S. L. Solheim. 1988. Forests too deer: edge effects in northern Wisconsin. *Conservation Biology* 2:348–58.

Arvard, M. S. 1993. Testing the "ecologically noble savage" hypothesis: interspecific prey choice by Piro hunters of Amazonian Peru. *Human Ecology* 21:355–87.

Atkinson, I. 1989. Introduced animals and extinctions. In *Conservation for the twenty-first century*, D. Western and M. Pearl (eds.), 54–69. New York: Oxford University Press.

Blaustein, A. R., and D. B. Wake. 1990. Declining amphibian populations: a global phenomenon? *Trends in Ecology and Evolution* 5:203–4.

Brown, L. 1994. Facing food insecurity. In *State of the world 1994*, L. R. Brown et al. (eds.), 177–97. New York: Norton.

Carlton, J. T., and J. B. Geller. 1993. Ecological roulette: the global transport of nonindigenous marine organisms. *Science* 261:78–82.

Diamond, J. M. 1984. "Normal" extinctions of isolated populations. In *Extinctions*, M. H. Nitecki (ed.), 193–246. Chicago: University of Chicago Press.

Dirzo, R. 1990. Contemporary neotropical defaunation and forest structure, function, and diversity—a sequel to John Terborgh. *Conservation Biology* 4:444–46.

Durning, A. T. 1993. *Saving the forests: what will it take?* Worldwatch Report No. 117. Washington, D.C.: Worldwatch Institute.

Ehrlich, P. R., and A. H. Ehrlich. 1981. *Extinction: the causes and consequences of the disappearance of species.* New York: Random House.

Fearnside, P. M. 1986. *Human carrying capacity of the Brazilian rainforest.* New York: Columbia University Press.

Gentry, A. H. 1986. Endemism in tropical versus temperate plant communities. In *Conservation biology: the science of scarcity and diversity*, M. E. Soulé (eds.), 153–81. Sunderland, Mass.: Sinauer.

Gilbert, L. E. 1980. Food web organization and the conservation of neotropical diversity. In *Conservation biology: an evolutionary-ecological perspective*, M. E. Soulé and B. A. Wilcox (eds.), 11–33. Sunderland, Mass.: Sinauer.

Goodland, R. J. A., and H. S. Irwin. 1975. *Amazon jungle: green hell to red desert?* Amsterdam: Elsevier.

Healy, R. G. 1985. *Competition for land in the American South: agriculture, human settlement, and the environment.* Washington, D.C.: Conservation Foundation.

Hedgpeth, J. W. 1993. Foreign invaders. *Science* 261:34–35.

Janzen, D. H. 1986. The external threat. In *Conservation biology: the science of scarcity and diversity*, M. E. Soulé (ed.), 286–308. Sunderland, Mass.: Sinauer.

Kaplan, R. 1994. The coming anarchy. *Atlantic Monthly* (Feb.): 44–76.

Lovejoy, T. E., R. O. Bierregaard Jr., A. B. Rylands, J. R. Malcolm, C. E. Quintela, L. H. Harper, K. S. Brown Jr., A. H. Powell, G. V. N. Powell, H. O. R. Schubart, and M. B. Hays. 1986. Edge and other effects of isolation on Amazon forest fragments. In *Conservation biology: the science of scarcity and diversity*, M. E. Soulé (ed.), 257–85. Sunderland, Mass.: Sinauer Associates.

Lowe-McConnell, R. H. 1987. *Ecological studies in tropical fish communities*. Cambridge: Cambridge University Press.

Martin, C. 1991. *The rainforests of West Africa; ecology, threats, conservation*. Basel: Birkhäuser.

Martin, P., and R. G. Klein, eds. 1984. *Quartenary extinctions: a prehistoric revolution*. Tucson: University of Arizona Press.

Miller, S. G., S. P. Bratton, and J. Hadidian. 1992. Impacts of white-tailed deer on endangered and threatened vascular plants. *Natural Areas Journal*. 12:67–74.

Myers, N. 1984. *The primary source: tropical forests and our future*. New York: Norton.

NRC [National Research Council]. 1980. *Conversion of tropical moist forests*. Washington, D.C.: National Academy of Sciences.

Olson, S. L., and H. F. James. 1982. Prodromus of the fossil avifauna of the Hawaiian Islands. *Smithsonian Contributions to Zoology* 365:1–59.

Peters, C. M., A. H. Gentry, and R. O. Mendelsohn. 1989. Valuation of an Amazonian rainforest. *Nature* 339:655–56.

Postel, S. 1994. Carrying capacity: Earth's bottom line. In *State of the world 1994*, L. R. Brown et al. (eds.), 3–21. New York: Norton.

Redford, K. H. 1992. The empty forest. *BioScience* 42:412–22.

Reid, W. V. 1992. How many species will there be? In *Tropical deforestation and species extinction*, T. C. Whitmore and J. A. Sayer (eds.), 55–73. London: Chapman and Hall.

Salafsky, N., B. L. Dugelby, and J. W. Terborgh. 1993. Can extractive reserves save the rain forest? An ecological and socioeconomic comparison of nontimber forest product extraction systems in Petén, Guatemala, and West Kalimantan, Indonesia. *Conservation Biology* 7:39–52.

Savidge, J. A. 1987. Extinction of an island forest avifauna by an introduced snake. *Ecology* 68:660–68.

Schonewald-Cox, C. M., S. M. Chambers, B. MacBryde, and L. Thomas, eds. 1983. *Genetics and conservation*. Menlo Park, Calif.: Benjamin Cummings.

Schwartzman, S. 1989. Extractive reserves: the rubber tappers' strategy for sustainable use of the Amazon rainforest. In *Fragile lands of Latin America*, J. O. Browder (ed.), 150–65. Boulder, Colo.: Westview Press.

Simberloff, D. S., and L. G. Abele. 1976. Island biogeography theory and conservation practice. *Science* 191:285–86.

Smith, F. D. M., R. M. May, R. Pellew, T. H. Johnson, and K. R. Walter. 1993. How much do we know about the current extinction rate? *Trends in Ecology and Evolution* 8:375–78.

Soulé, M. E., ed. 1987. *Viable populations for conservation*. Cambridge: Cambridge University Press.

Soulé, M. E., E. T. Bolger, A. C. Alberts, J. Wright, M. Sorice, and S. Hill. 1988. Reconstructed dynamics of rapid extinctions of chaparral-requiring birds in urban habitat islands. *Conservation Biology* 2:75–92.

Temple, S. A. 1977. Plant-animal mutualism: coevolution with dodo leads to near extinction of plant. *Science* 197:885–86.

Terborgh, J. 1988. The big things that run the world—a sequel to E. O. Wilson. *Conservation Biology* 2:402–3.

Terborgh, J. 1992a. *Tropical deforestation.* Burlington, N.C.: Carolina Biological Supply.

Terborgh, J. 1992b. Maintenance of diversity in tropical forests. *Biotropica* 24:283–92.

Terborgh, J. 1992c. *Diversity and the tropical rain forest.* New York: Freeman.

Terborgh, J., and B. Winter. 1980. Some causes of extinction. In *Conservation biology: an evolutionary-ecological perspective,* M. E. Soulé and B. A. Wilcox (eds.), 119–33. Sunderland, Mass.: Sinauer.

Tutin, C. E. G., E. A. Williamson, M. E. Rogers, and M. Fernandez. 1991. A case study of plant-animal relationships: *Cola lizae* and lowland gorillas in the Lopé Reserve, Gabon. *Journal of Tropical Ecology* 7:181–99.

Urban, F., and R. Nightingale. 1993. *World population by country and region, 1950–90, and projections to 2050.* Washington, D.C.: Economic Research Service, U.S. Bureau of the Census, U.S. Department of Agriculture.

Warner, R. E. 1968. The role of introduced diseases in the extinction of the endemic Hawaiian avifauna. *Condor* 70:101–20.

WCMC [World Conservation Monitoring Centre]. 1992. *Global biodiversity: status of the earth's living resources.* London: Chapman and Hall.

Whitmore, T. C. 1990. *An introduction to tropical rain forests.* Oxford: Clarendon Press.

Whitmore, T. C., and J. A. Sayer. 1992. *Tropical deforestation and species extinction.* London: Chapman and Hall.

Wilcox, B. A. 1980. Insular ecology and conservation. In *Conservation biology: an evolutionary-ecological perspective,* M. E. Soulé and B. A. Wilcox (eds.), 95–117. Sunderland, Mass.: Sinauer.

Wilson, E. O. 1992. *The diversity of life.* Cambridge, Mass.: Harvard University Press.

Zaret, T. M. 1980. *Predation and freshwater communities.* New Haven, Conn.: Yale University Press.

Zaret, T. M., and R. T. Paine. 1973. Species introduction in a tropical lake. *Science* 182:449–55.

# 3

# The Ecological Foundations of Biodiversity Protection

*Kathy MacKinnon*

Two thirds of all known species occur in tropical regions, and probably half of all species are confined to tropical rain forests—yet these rain forests are among the most threatened of all natural habitats. Throughout the tropics, rain forests are being destroyed at an alarming rate. It has been estimated that, worldwide, approximately 170,000 square kilometers of rain forests—an area almost as great as Cambodia—are being lost every year (FAO, 1990).

Few tropical countries retain more than half of their natural forest cover, and even those that do are witnessing rapid habitat conversion. Figure 3-1 illustrates the decline of primary forest cover on Sumatra during the past 60 years, a picture that is duplicated over much of tropical Asia. The Indonesian archipelago, as a whole, loses at least 9,000 square kilometers of forest each year to logging, land conversion, and shifting agriculture (MoF/FAO, 1991). In some years, the figure is even higher. In 1982 and 1983, for example, severe drought and fires (often deliberately started) damaged 36,000 square kilometers of forest—an area the size of Belgium—in East Kalimantan in Indonesian Borneo (Lennertz and Panzer, 1983) and another 10,000 square kilometers in Sabah (Malingreau et al., 1985).

Destruction of tropical habitats leads to the irreversible loss of biological diversity and genetic resources. Conservation of biodiversity will require a concerted effort to provide adequate and effective protection of tropical forests and their native species. The best, easiest, and least expensive way to achieve this goal is to establish networks of protected rain forest areas for in situ conservation of gene pools, species, and ecosystems. Forest destruction is proceeding so fast that this decade is probably the last chance to protect extensive areas of tropical forests; indeed, for some countries it is already too late.

While this chapter focuses primarily on tropical Asia, many of the lessons and recommendations apply equally well to the rain forests of Africa and Central and South America.

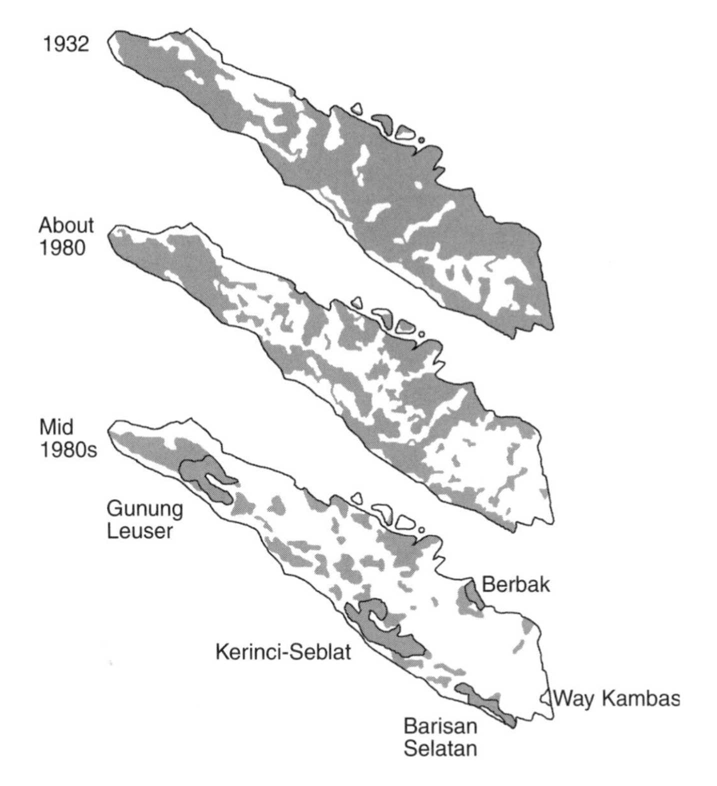


***Figure 3-1*** Loss of primary forests in Sumatra in recent years, showing areas of national parks on latest map (after Whitten et al., 1987).

## PROTECTING HABITATS: HOW MUCH IS ENOUGH?

The question of how much protected habitat is enough has long been debated by conservationists and other scientists. How much forest must be reserved to protect all species and all habitats? How many reserves are needed? Where should those reserves be located? And how large should reserves be? Taking into account other demands on land for subsistence and economic development, the World Conservation Union (IUCN) has recommended that nations allocate 10 percent of their land for conservation areas (McNeely and Miller, 1983). Few tropical countries have achieved this target. In Asia only Bhutan, Brunei, and Indonesia have so far approached this figure, and many so-called protected areas exist on paper only (MacKinnon and MacKinnon, 1986a). In Africa several nations, including Benin, Senegal, Tanzania, Zimbabwe, and the Central African Republic, have officially protected more than 10 percent of their land, but this protection is primarily in savanna areas rather than in tropical rain forests (MacKinnon and MacKinnon, 1986b).

Even a protected area network that covers 10 percent of a nation's forests may be inadequate to protect the country's biodiversity. Studies of protected areas throughout Asia and Africa show that while some habitats such as montane forests are generally well protected, species-rich lowland forests and wetlands, including coastal mangroves, are often poorly represented (MacKinnon and MacKinnon, 1986a, 1986b). Thus, only 2.9 percent of Bornean mixed lowland rain forest and less than 1 percent of the island's mangrove forests are designated as reserves, compared with 11 percent of montane forests, which are less accessible for agriculture and exploitation.

## THE IMPORTANCE OF LOWLAND RAIN FORESTS

The greatest richness of plant and animal species is concentrated in lowland forests. More than half the mammals of peninsular Malaysia do not occur above an altitude of 350 meters, and 81 percent are restricted to altitudes below 660 meters (Stevens, 1968). This makes protected areas in lowland forests of particular importance for biodiversity conservation. The Taman Negara conservation area (4,643 square kilometers in size) in West Malaysia encompasses a large block of lowland rain forests as well as hill and montane rain forests, and it supports 60 percent of the endemic mammal species of the entire Sunda Shelf region. Of the 198 mammals recorded in the conservation area, 142 (71 percent) are dependent on the rain forests (Medway, 1971). Of the 241 lowland bird species recorded for Peninsular Malaysia, 172 occur in the Taman Negara (Wells, 1971). The situation is similar on Borneo, where 61 percent of all resident birds are confined to mixed lowland forests; 60 percent of these birds (146 species) are Sunda endemics (Wells, 1984). Mammal and bird lists for Bornean reserves confirm the richness of lowland forest habitats. For example, 274 birds, half the Bornean list, occur in the lowland forests of Kutai National Park in East Kalimantan (MacKinnon et al., 1996).

Lowland forests are also important as a colonizing source for higher elevations. This is well illustrated by plotting the altitudinal distribution of forest birds on the Greater Sunda islands, as shown in Figure 3-2. The Borneo and Sumatra graphs show maximum bird richness in the lowland forests and a gradual reduction in numbers of species with increasing altitude, which is only slightly compensated for by the appearance of montane species at higher altitudes. This pattern is typical of other tropical regions, including New Guinea. The Java graph is atypical in having fewer species in the hill zone between 300 meters and 1,000 meters than at altitudes between 1,000 meters and 1,500 meters. This is probably due to the loss of some lowland species normally found at middle altitudes as a result of being cut off from lowland sources of recolonizers (MacKinnon and Phillipps, 1993).

Lowland forests are obviously of crucial importance for conservation of biodiversity. Yet these are some of the most accessible and most productive habitats and therefore some of the most threatened by logging, agriculture, and industrial development.

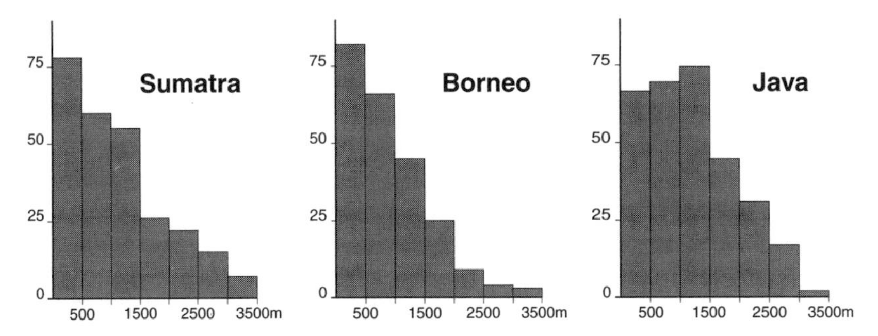


*Figure 3-2* Proportions of island bird fauna at different altitudes on Sumatra, Borneo, and Java, showing reduction in middle-level species on Java because of isolation from colonizing populations in lowland forests. *Source:* MacKinnon and Phillipps, 1993.

## PLANNING A PROTECTED AREA NETWORK

While species richness is a characteristic of mixed lowland rain forests, endemism is more often associated with islands and mountains. On Borneo, for example, most endemic birds and mammals occur in montane habitats. Since almost all habitat types contain some unique species or species communities, an effective protected area network must protect representative samples of all major habitat types in all biogeographic units. New reserves should be established to complement those already in the network and to bring more habitats and species under protection. Protecting the entire range of habitat types and ecosystems will not only ensure the conservation of the widest possible range of species; it should also help to ensure long-term survival of species if climatic conditions and habitat distributions change due to global warming.

Biological diversity is not distributed uniformly: centers of biological richness or endemism can be recognized. As shown in Figure 3-3, sites and habitats that are important for conservation of one animal group, such as African primates, may also be important for other animal groups and plants (MacKinnon and MacKinnon, 1986b). This means that better known animal groups, such as birds, may be useful indicators for identifying areas of high biodiversity value. There is considerable overlap between areas of endemism for birds, amphibians, and reptiles in Central America, and between areas of endemism for birds, amphibians, and mammals in Africa (Bibby et al., 1992). In Indonesia, gazetting just six of the new reserves proposed for the Moluccas in the country's National Conservation Plan (MacKinnon and Artha, 1982) would protect more than 120 species of restricted-range and endemic birds.

Although most tropical countries would benefit from a more detailed inventory of tropical forests, this should not be taken as an argument for delaying the planning and establishment of conservation areas.

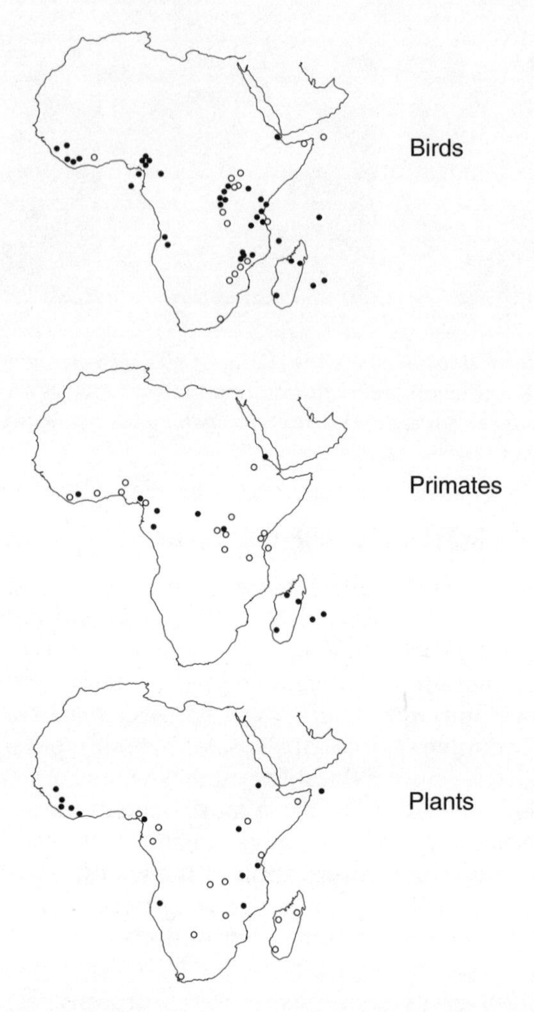


*Figure 3-3* Sites of primary (•) and secondary (o) conservation importance for birds, primates, and plants in Africa. *Source:* MacKinnon and MacKinnon, 1986b.

For most countries, sufficient information already exists to ascertain areas of high biodiversity and conservation priority and to plan effective protected area networks. The fact that much of this information is often in foreign museums and herbaria emphasizes the need for greater international cooperation and freer access and exchange of information to aid biodiversity conservation.

The Conservation Needs Assessment (CNA) project in Papua New Guinea is a good example of an exercise to compile and synthesize large quantities of geographical and distributional data relevant to biodiversity conservation. By plotting the overlap of areas important for different plant and animal groups, the CNA project was able to identify more than

40 key areas for biodiversity conservation (Beehler, 1993). Rapid appraisal techniques can provide a scientific basis for selecting and establishing networks of protected areas (Margules et al., 1988, 1994), as well as for establishing baseline data for monitoring and evaluating the effectiveness of management once protected areas are established.

Since individuals of any one species tend to be rare in rain forests, the area required to protect a minimum viable population will be large. How large will depend on the species of animal or plant to be conserved and its distribution or ranging patterns. Davies and Payne (1982) estimated the minimum contiguous area of Borneo rain forest required to conserve a population of 200 adults of various mammal and bird species. For large mammals, such as elephant and rhino, an area of 6,000 square kilometers (600,000 hectares) may be needed. Wide-ranging species such as the tiger and other large cats found in Sumatra and mainland Asia may require even larger areas. Moreover, a population of 200 adults may be too small to survive over the long term, since not all of the individuals will necessarily contribute to the breeding pool. Even a breeding population of hornbills is likely to require 4,000 square kilometers. The message is clear: rain forest reserves need to be large. Yet few gazetted rain forest reserves exceed 6,000 square kilometers, and many are much smaller.

Even a large block of habitat is likely to lose some species if the area becomes cut off from other areas of similar natural habitat (see chapter 2). Several classic studies of the biogeography of islands (Diamond, 1975) have shown that small islands are unable to support as many species as larger islands of similar habitat. Species numbers are the result of a balance between species immigration and extinction, determined by island size and distance from the mainland. Where conservation areas become "habitat islands," they will lose some of their original species until a new equilibrium is reached, which will depend on the size of the area, its richness and diversity of species, and its distance from a colonizing source. As a rough generalization, a reserve encompassing just 10 percent of the original habitat will retain only 50 percent of the original species present. Larger patches of habitat will lose species more slowly but any loss of habitat will lead to some loss of species. In very small fragments of habitat, not only will there be species loss related to the area's size, but there will also be additional, and often rapid, reduction in remaining species due to secondary extinctions (see chapter 2).

Small areas of forest lose proportionally far more species than larger blocks. On Java, small forest patches 10–40 hectares in size have lost up to 80 percent of their expected original bird species, compared to only a 25 percent loss in forest areas of more than 10,000 hectares (van Helvoort, 1981). One well-documented area is the Bogor Botanic Gardens. In 1947, 142 species of birds were recorded in the gardens, with 100 of them being regular visitors (Hoogerwerf, 1949). During the past 20 years, only 81 of the original species have been recorded (a 46 percent loss of species), and today only 40 species are regularly observed (van Balen et al., 1988). The habitat in the gardens has not deteriorated, but land use in the surrounding countryside has changed so that the gardens are now far from any blocks of natural forest that could serve as recolonizing sources. These results have important implications for conservation. Java and Bali, for example, have more than 100 conservation areas, but most of them are very small and isolated from other blocks of natural forest. Only the 12 reserves larger than 100 square kilometers are likely to retain anything like their original complement of bird species.

It is worth noting, however, that the distance between forest fragments is also an important deciding factor in rate of species loss. Thus, fragments of forest around Lake Victoria in Uganda are "behaving" like parts of a larger contiguous forest and are losing species much more slowly than would be expected for their size. The forest fragments are still sufficiently close that birds, butterflies, bats, and some other mammals are able to traverse the distances between them (P. Howard, pers. comm., 1994). This has important implications for conservation planning. Maintaining corridors of natural habitat between reserves and neighboring forests not only increases the overall area of habitat for wildlife but also provides opportunities for gene flow and migration between populations.

An optimal conservation strategy would be to establish a national network of protected areas to represent all major habitat types within each biogeographic zone, with priority areas complementing one another for habitat and species coverage. Indonesia is one of the few Asian countries to have prepared such a system plan (FAO, 1981–1982). Within each biogeographic region, the main priority was to establish one or more large reserves (100,000 hectares or more) selected to include a continuum of habitat types. These reserves are augmented with smaller reserves that protect additional habitat types or sites of special interest. Priority areas for conservation were identified according to size, biological value (species richness, endemism, habitat variety, and extent), socioeconomic value (e.g., watershed protection), and manageability (e.g., lack of conflict with other land uses). The Indonesian National Conservation Plan identified more than 80 priority areas for conservation and established national priorities for funding and development, as shown in Figure 3-4.

Similar conservation planning could, and should, be undertaken in other tropical countries with large areas of natural habitats remaining, such as Laos, Cambodia, Myanmar, and Papua New Guinea. In many tropical countries, however, deforestation is already so extensive that few conservation options remain and the debate on reserve size and design is no longer relevant. Thailand, for instance, has less than 20 percent of forest cover remaining (much of it already protected) and needs to secure its existing reserves, particularly the large western for-

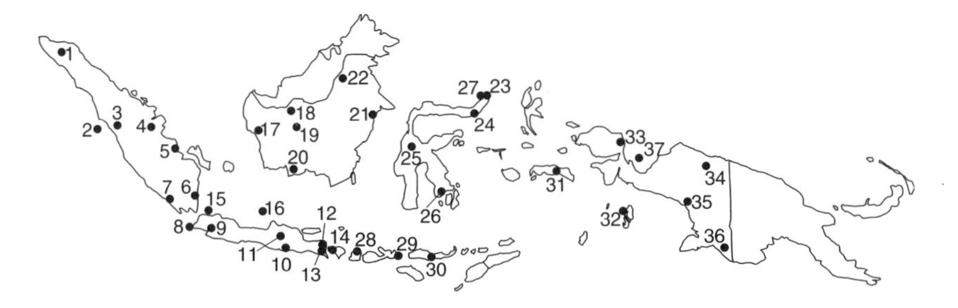

*Figure 3-4* National parks and other major protected areas under development in Indonesia. NP = national park; m = marine. 1. Gunung Leuser NP; 2. Taitai Batti, Siberut; 3. Kerinci-Seblat NP; 4. Seberida; 5. Berbak NP; 6. Way Kambas NP; 7. Barisan Selatan NP; 8. Ujung Kulon; 9. Gunung Gede-Pangrango; 10. Meru Betiri; 11. Bromo; 12. Baluran; 13. Alas Purwo; 14. Bali Barat; 15. Pulau Seribu (m); 16. Karimunjawa (m); 17. Gunung Palung; 18. Danau Sentarum; 19. Bukit Baka; 20. Tanjung Puting; 21. Kutai; 22. Kayan-Mentarang; 23. Tangkoko; 24. Dumoga-Bone; 25. Lore Lindu; 26. Rawa Aopa; 27. Manado (m); 28. Mt. Rinjani; 29. Komodo; 30. Ruteng; 31. Manusela; 32. Aru (m); 33. Arfak; 34. Mamberamo; 35. Gunung Lorentz; 36. Wasur NP; 37. Teluk Cendrawasih (m).

est complex around Huai Kha Khaeng and Thung Yai, which includes much of the country's rich biodiversity (Nakasathien and Stewart-Cox, 1990).

Underlying the emphasis on protecting large blocks of habitat is the assumption that if adequate habitat is protected, the constituent species will also be protected. Consider the situation on Borneo. Graphs that plot the distribution of Bornean birds, endemic and threatened mammals, and other indicator species show that almost all are represented in at least one of the island's conservation areas and many species occur in several areas, as shown in Figure 3-5. Other country and regional reviews give a similar picture (MacKinnon and Wind, 1980). In Thailand, where the protected area network encompasses only a few large reserves and many small reserves, many established on an ad hoc basis, most conspicuous mammals and birds and endemic vertebrates are recorded from existing reserves. A detailed analysis showed that 508 of Thailand's 578 resident forest birds are recorded from protected areas and more are likely to be found as surveys continue (Round, 1988). Even in Indochina, where protected areas are among the least adequate in Asia, most large mammals are recorded from at least one reserve (MacKinnon and MacKinnon, 1986a). However, their long-term survival will depend on effective protection and management of those reserves, and on the nearness and linkages of the reserves to other areas of undisturbed natural habitat.

Of course, the presence of a species in a reserve at any particular moment may be no guarantee for its long-term survival, especially since

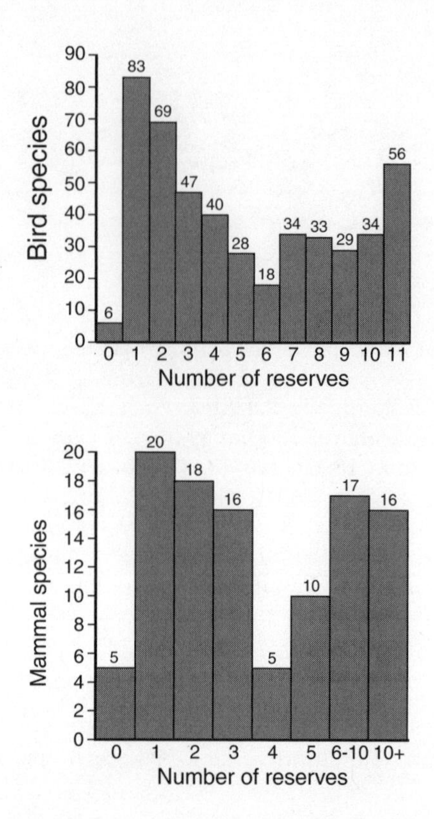


***Figure 3-5*** Incidence of selected bird and mammal species in Bornean reserves. Note that almost all species are recorded in at least one reserve. Forty-nine percent of all birds are recorded in five or more reserves, and 45 percent of all mammals are recorded in four or more reserves. *Source:* MacKinnon et al., 1996.

rain forest species are rare and occur at low densities. Table 3-1 illustrates areas of remaining and protected habitat for the primate species of the Indomalayan realm. Only seven species—the white-handed gibbon (*Hylobates lar*), the dusky langur (*Presbytis obscura*), the Sulawesi macaque (*Macaca nigra*), and the four endemic Mentawai primates—have 10 percent or more of their habitat protected within reserves, and five of these species have extremely limited distributions. Whereas the more widely distributed gibbon and dusky langur occur in several reserves in two or three countries, the Mentawai primates are confined to the Mentawai islands, are secure only in the Taitai Batti reserve, and are threatened by logging operations, overhunting, and new plans for oil-palm plantations on Siberut. Thus, even species with 10 percent or more of their habitat protected are vulnerable if their distribution is limited.

***Table 3-1*** Range loss and habitat protected for primates in Southeast Asia

| Species | Original habitat (km²) | Remaining habitat(km²) | Percentage loss | Percentage protected |
|---|---|---|---|---|
| *Pongo pygmaeus* | 333,000 | 207,000 | 63 | 2.1 |
| *Hylobates syndactylus* | 465,110 | 169,800 | 63 | 6.8 |
| *H. agilis* | 532,270 | 184,345 | 65 | 3.7 |
| *H. lar* | 280,700 | 100,240 | 64 | 13.5[a] |
| *H. muelleri* | 395,000 | 253,000 | 36 | 5.1 |
| *H. klossii* | 6,500 | 4,500 | 31 | 22.9[a] |
| *H. moloch* | 43,274 | 1,608 | 96 | 1.3 |
| *H. concolor* | 349,330 | 87,532 | 75 | 3.1 |
| *H. hoolock* | 168,353 | 56,378 | 67 | 5.1 |
| *H. pileatus* | 70,000 | 11,200 | 84 | 9.9 |
| *Macaca fascicularis* | 383,181 | 123,315 | 68 | 3.4 |
| *M. nemestrina* | 1,568,623 | 481,685 | 69 | 4.1 |
| *M. arctoides* | 1,546,964 | 556,466 | 64 | 3.7 |
| *M. assamensis* | 802,193 | 335,002 | 59 | 2.5 |
| *M. mulatta* | 1,732,270 | 568,638 | 67 | 2.8 |
| *M. pagenis* | 6,500 | 4,500 | 31 | 22.9[a] |
| *M. nigra/nigrescens* | 12,000 | 4,800 | 60 | 22.9[a] |
| *M. tokeana/hecki* | 67,000 | 38,500 | 33 | 1.5 |
| *M. maura* | 23,000 | 2,800 | 88 | 2.1 |
| *M. ochreata/brunnescens* | 29,500 | 18,500 | 37 | 4.8 |
| *Nasalis larvatus* | 29,496 | 17,750 | 40 | 4.1 |
| *Simias concolor* | 6,500 | 4,500 | 31 | 22.9[a] |
| *Pygathrix avunculus* | 29,688 | 9,060 | 70 | 1.5 |
| *P. nemaeus* | 296,000 | 72,270 | 76 | 3.1 |
| *Presbytis comata* | 43,274 | 1,608 | 96 | 1.6 |
| *P. frontata* | 125,000 | 61,500 | 51 | 5.8 |
| *P. hosei* | 104,000 | 54,000 | 48 | 6.8 |
| *P. melalophos/femoralis* | 450,834 | 168,227 | 63 | 7.5 |
| *P. obscura* | 155,494 | 56,767 | 64 | 15.1[a] |
| *P. rubicunda* | 415,000 | 266,000 | 36 | 4.7 |
| *P. potenziani* | 6,500 | 4,500 | 31 | 22.9[a] |
| *P. thomasi* | 68,000 | 30,800 | 55 | 5.8 |
| *Trachypithecus auratus* | 46,890 | 6,898 | 86 | 1.7 |
| *T. cristatus* | 412,170 | 169,970 | 59 | 3.9 |
| *T. francoisi* | 97,400 | 14,106 | 86 | 1.2 |
| *T. phayrei* | 708,572 | 193,192 | 73 | 3.8 |
| *T. pileatus* | 65,868 | 34,622 | 48 | 5.6 |
| *Tarsius bencanus* | 609,000 | 305,000 | 50 | 5.4 |
| *T. spectrum* | 154,400 | 70,730 | 54 | 3.8 |
| *T. syrichta* | 60,248 | 13,050 | 79 | 0.8 |
| *Nycticebus coucang* | 2,109,700 | 653,545 | 69 | 4.4 |
| *N. pygmaeus* | 296,000 | 72,720 | 76 | 3.1 |

*Sources:* MacKinnon, 1986; MacKinnon and MacKinnon, 1986a, 1987.
[a]More than 10 percent of habitat protected.

## INTEGRATING PROTECTED AREAS WITH NATIONAL AND LOCAL DEVELOPMENT

While the most important areas for biodiversity conservation must be identified according to biological criteria, the long-term survival, protection, and management of those areas will depend on a host of other political, social, and economic factors as well. Declaring an area protected, marking the boundary, and employing guards are all essential steps, but they are unlikely to achieve effective protection without strong commitment and support from local communities, regional and national governments, and all concerned agencies. Many of the practical aspects of achieving such commitments from the various stakeholders are reviewed elsewhere in this book, particularly in chapters 5 and 9.

Gaining widespread support for conservation efforts often requires changing the public's attitude to recognize that conservation areas can be a legitimate and wise form of land use. The ecological and environmental benefits of protected areas and natural forests are almost never fully recognized or acknowledged, yet such benefits as the protection of watersheds and soils and the conservation of genetic resources may far outweigh the "costs" in lost development opportunities (MacKinnon et al., 1986; McNeely, 1988). In fact, the watershed-protection argument may offer one of the most effective means of gaining support for protected areas; nearly everyone understands the need for water, whereas few nonscientists understand the need for biodiversity conservation and most people tend to see it as an "animals versus people" controversy.

Another important requirement for forging successful conservation efforts is to convince politicians and development agencies of a key tenet held by most conservationists—that conservation of biodiversity is fundamental to any form of environmentally sustainable development. It is also true that conserving biodiversity requires wise management of the entire landscape, from strictly protected areas to exploited natural habitats (where the aim should be to implement sustainable harvesting regimes) and intensively managed agricultural lands.

Recognizing that conservation areas are a rational form of land use is the first step toward integrating the areas into national and regional development programs. Although this remains a largely unfulfilled goal, several successes are notable. For example, Costa Rica has adopted a new strategy that allocates 20 percent of the country's land area for biodiversity (conservation areas) and 80 percent for agriculture, making it perhaps the first tropical country to fully recognize the true value of biodiversity goods and services in regional and national accounting.

Indonesia offers another example. Establishment and protection of the Dumoga-Bone National Park in North Sulawesi was considered crucial to the success of a large-scale irrigation project designed to increase the agricultural potential of the Dumoga valley. The park protects the forested watershed—as well as the $54 million investment by the World

Bank in the irrigation project—at a cost of less than 1 percent of the total project. The park not only conserves biodiversity but also helps to minimize siltation in vital waterways and to ensure a steady year-round flow of water (Sumardja et al., 1984).

Unfortunately, the Dumoga-Bone story is one of the few examples of a relatively happy "marriage" between conservation and development. More often, development activities can and do threaten protected areas and natural habitats. For example, the creation of networks of roads, either to extend national transportation systems or to serve logging activities, provides access to previously remote areas for pioneer farmers, cattle ranchers, and miners. New roads also offer ready access to distant markets, which may encourage previously remote communities to increase their forest harvesting and swidden (formerly called slash-and-burn) agriculture to levels that are no longer sustainable. While the actual building of roads may destroy only small areas of forest, the subsequent clearing of forest along the roads can be dramatic and devastating, as can be witnessed along the trans-Amazonia highway. Indeed, the effects of road development on deforestation and forest fragmentation can be so serious that it may often be worth considering rerouting roads to avoid areas of high biodiversity value and undisturbed forest. One study in the Amazon, for example, has shown that the additional costs of rerouting a new section of highway would be more than compensated by preserving the pristine forest's biodiversity and its "carbon sink values" (J. Spears, pers. comm., 1994).

Special caution needs to be taken when considering the need for roads in and around protected areas (MacKinnon et al., 1986; Peres and Terborgh, 1994). Some roads may be necessary for marking boundaries, for patrolling the area, and to support ecotourism. However, unless the parks also have strong law enforcement, effective management, and well-trained and well-equipped staff, such access roads can exacerbate incursion problems and lead to further destruction of habitats and wildlife.

## INTEGRATING PROTECTED AREAS WITH REGIONAL DEVELOPMENT

One recent concept being applied for integrating protected areas with regional development is the use of so-called integrated conservation-development projects (ICDPs) (Wells and Brandon, 1992). Most ICDPs have been small, local efforts to relieve pressure on parks and reserves by encouraging small-scale alternative economic activities. Some, however, have been more ambitious. For example, the ICDP proposed for Kerinci-Seblat National Park in Indonesia will require not only support from local communities but also from local, provincial, and national planners so that the conservation area is an integrated and valued part of regional development.

The Kerinci-Seblat National Park (KSNP) in Sumatra is one of the largest and most important conservation areas in Southeast Asia. Extending along the Barisan range, the park straddles four provinces and covers almost 1 million hectares. The park and adjacent forests encompass a range of habitats from species-rich lowland forests through hill forests and unique highland wetland systems to montane forests and subalpine habitats on Mt. Kerinci, at 3,805 meters Sumatra's highest mountain. The park is remarkable for its species richness, with records of more than 4,000 plants (1/60th of the world's total); 180 birds (1/50th of all birds), including 14 of the 20 Sumatran mainland endemics; and 144 mammals (73 percent of the Sumatran mammal fauna and 1/30th of the world's total). This area harbors some of the last viable populations of rare and endangered mammals, such as the endemic Sumatran hare (*Nesolagus netscheri*), the Sumatran rhinoceros (*Dicerorhinus sumatrensis*), the clouded leopard (*Neofelis nebulosa*), the Sumatran tiger (*Panthera tigris*), the Malay tapir (*Tapirus indicus*), and the Asian elephant (*Elephas maximus*). Many of the wide-ranging large herbivores and predators require large areas of lowland forests and other natural habitats to provide adequate feeding grounds and access to vital mineral licks. In addition to its biological values, the park protects the watersheds of the Musi and Batang Hari rivers, thus safeguarding water supplies to an estimated 3.5 million people and 7 million hectares of agricultural land.

Since late 1991 the World Bank has been assisting the government of Indonesia to prepare an ICDP for Kerinci-Seblat, to be funded partly through a grant from the Global Environment Facility (GEF) and partly through a Bank loan. KSNP is an ideal candidate for GEF funding since it supports biodiversity of global significance but is beset by many of the land-use problems that face other parks in densely populated Asia. Large areas of lowland rain forests have already been lost to logging (both government sanctioned and illegal) and conversion to plantations and agriculture. In the south, commercial mineral exploration and artisanal gold mining also threaten the park. Its long, narrow shape makes the park particularly vulnerable to fragmentation.

The park overlaps four provinces and nine regencies (kabupatens), and until recently there had been no attempt to coordinate regional development efforts to relieve pressure on the park. In 1991 the national planning board BAPPENAS established a regional coordinating committee involving all local government agencies to address the future development of the park and its environs. This committee supports the ICDP process, although it is not clear whether development projects undertaken to date (40 percent of them bridge and road building) will draw populations away from the park or actually open the hinterland to more developments that further threaten the park's integrity.

The park's location and the nature of the local economy present formidable obstacles to its conservation. KSNP is surrounded by areas of high population density, and large areas within the park have been cleared of forest, both for subsistence farming and for cassiavera

(*Cinnamomum burmanni*) production. In and around the Kerinci valley, many farmers are engaged in cultivation of cassiavera, a highly profitable cash crop, which has made many valley residents wealthy. Many of the farmers engaged in this activity are not subsistence farmers or even long-time residents of the park; many hold other land outside park boundaries and are often sponsored by rich patrons who subsidize the enterprise and receive a percentage of the crop. From the patrons' point of view, investment in cinnamon production brings a very high return. Until recently, neither park authorities nor local government has attempted to address this problem; indeed, many high officials and community leaders have cassiavera holdings.

While it is probably true that a "traditional" policing approach alone would not have protected KSNP from encroachment, in fact it has never been tried. The park is grossly understaffed, underbudgeted, and generally under- or even unresourced. From 1982 until recently, the park was run as a project activity with little money or manpower. In 1991, when the Indonesian government began its ICDP approach, the annual budget was increased from $75,000 to $550,000; in 1992, this was raised to $700,000, an investment of 70 cents per hectare. The park management authority (PHPA) has only 75 staff, many untrained, with half of them based in the Kerinci valley. With such meager resources PHPA exercises no real control over the park.

Since 1982, when the park was first proposed, almost all lowland forests and accessible hill forests within the original park boundaries have been excised as logging concessions, even though large areas within some concessions are officially designated as protection forests to protect watershed functions. Since the park's high biodiversity values are dependent on maintaining a continuum of habitats from lowland forests to subalpine montane systems, it is imperative to conserve forests and wildlife in the concession buffer zone surrounding the park as well as in the park itself.

The GEF project aims to secure the biodiversity of KSNP and stop further habitat fragmentation, by improving protection and management of the park and by promoting sustainable management and the maintenance of permanent forest cover in the buffer-zone logging concessions. The project will develop a model for integrating conservation and regional and district development, which can be applied to other parks in Indonesia and elsewhere in Asia. This integrated approach will help stabilize the park boundary and reduce pressure on biodiversity-rich lands by providing local communities with alternative livelihood opportunities and rural development consistent with park conservation objectives. A particular concern is to minimize the impact of logging in the species-rich lowland and lower-hill forests and to ensure that lightly logged forest is not then cleared for agriculture but is maintained as a permanent forest buffer zone (World Bank, 1995).

As part of forest-management activities, the Indonesian Ministry of Forestry will review and revise the boundary between the park and the

17 neighboring concessions according to criteria based on biodiversity, land use, and biophysical and watershed values. Rapid ecological assessment surveys will determine areas of high biodiversity within the concessions. Areas identified as important for biodiversity or watershed conservation will be returned to the park or maintained as protection forest within the concessions. Technical assistance, training, monitoring, and independent auditing will help concessionaires implement sustainable management regimes and maintain permanent forest cover after logging. Some logged areas will be allocated as community forests, to be managed by local communities. Improved forestry management for conservation and sustainable use in the Kerinci forest buffer zone will not only effectively increase the conservation estate, by maintaining natural habitat beyond park boundaries, but could provide a model for sustainable forestry elsewhere in Indonesia.

A regional impact assessment (RIA) identified four major threats to biodiversity: roads (which open new areas, thereby leading to fragmentation), mining (industrial companies and artisanal gold miners), loss of lowland forest habitat, and agricultural encroachment. As a result of the RIA, the government has agreed to postpone any new road development until after the preparation of a regional spatial development plan and park management plan, which will include recommendations for zoning and protecting areas of high biodiversity. Similarly, while the three mining concessions that overlap park boundaries will be allowed to continue their exploration activities, no further concessions will be granted until park boundaries are agreed. Any proposals to mine mineral deposits will be evaluated on a case-by-case basis through a rigorous environmental impact process, with the added recommendation that any mining infrastructure would be outside park boundaries.

The long-term survival of KSNP will depend on strong political will, on better appreciation of the park's ecological and environmental values, and, crucially, on how the social and economic issues are resolved. Strict law enforcement will be necessary to protect the area's habitats and stop fragmentation, but enforcement alone cannot save the park. There will also need to be close consultation with and involvement of local communities in planning and management to engage their support and to provide alternative and appropriate income-generating opportunities that safeguard park benefits. Socioeconomic surveys are already being conducted to determine the status of park residents, landholders, and encroachers. It is imperative to know who is destroying park resources and why, in order to know how best to reconcile the legitimate needs of local communities with conservation. Individual cases will require individual solutions, but it is certain that some of those currently farming within the park will have to stop.

Intensification of agriculture is an obvious solution for buffer communities, but it may be impossible to find a crop as profitable as cassiavera, given that the cassiavera growers are effectively mining the fertility of the park's forests without bearing any of the real costs of production. It is ironic that while much attention has been given to the rights of settlers (many recent, sponsored immigrants) to the east of the park, indigenous Kubu people (true forest dwellers and hunter gatherers) have been made homeless by a World Bank–funded transmigration scheme. In fact, the Kubu could benefit from conservation of rain forests in the park and in logging concessions.

The issues in and around KSNP are complex. Dealing with them will require considerable patience, negotiation, compromise, and good collaboration between different sectors, government agencies, local communities, and international and national nongovernmental organizations. The park's history has been marked by a confusion of conflicting agendas and objectives, with one government agency "giving away" areas that another agency, under the same minister, is pledged to protect. There has been conflict between the objectives of different local development agencies and between local government and decision makers and planners in the capital of Jakarta. What is certain is that if agricultural encroachment, legal and illegal logging, gold mining, and inappropriate developments are not stopped, then very soon the park will become fragmented into a series of "stepping stones" of highland habitats. The area's biological value will be much reduced. The best hope for the future of KSNP is the development of a fully integrated conservation and management strategy for the park and its environs, developed in the context of regional development plans and involving all intersectoral agencies. Not only will this allow sensible regional planning that enhances and builds on the park's important environmental roles, but existing government development budgets can then be reallocated to fund developments that maintain, rather than destroy, the biological values of the park. The GEF project is a major opportunity to protect KSNP.

## INVOLVING LOCAL COMMUNITIES IN PROTECTED AREA MANAGEMENT

Experience teaches that conservation measures must take into account the needs and aspirations of local peoples and the established patterns of local land tenure. Even with government commitment to conservation, protected area management is unlikely to succeed without the participation and active support of local people and communities. Some of the people may have derived much or all of their livelihood from the designated conservation area and thus may be likely to see their rights curtailed; others may be recent immigrants exploiting forest resources and clearing land for agriculture. In both cases, their legitimate needs for land and livelihoods must be addressed. Reconciling the needs of conservation and local communities is a complex and difficult task, whether at the local level or in the regional context.

The task is made more difficult by the widely promoted myth that indigenous peoples always, or even usually, live in harmony with their natural environments and are wise custodians of the world's biodiversity. There are, indeed, areas in the world where native peoples, following traditional cultural practices on their own lands, protect large areas of essentially natural habitats while sustainably harvesting natural resources (McNeely and Pitt, 1985). More often, however, reality fails to approach the ideal, especially where indigenous populations are increasing and the people are gaining easier access to modern technology and widespread markets. Even where indigenous peoples have been given custody over their traditional lands, not all of them are harvesting the forests sustainably. For example, the Kayapo Indians of eastern Amazonia are estimated to have earned $33 million from felling mahogany in their indigenous reserve in 1988. These levels of harvesting are unsustainable but likely to continue, given the Brazilian government's inability to pay each Indian community $50,000 a month in requested compensation to stop the logging (Peres, 1994).

Compounding matters, development policies and in-country emigration have created numerous situations in which communities within or adjacent to parks comprise both indigenous people and pioneer settlers who arrived more recently. Often, the new arrivals are responsible for more destructive land-use practices than are the indigenous people, but even the latter may change their land-use practices as new technologies and markets become available to them (Kartawinata et al., 1984).

It is clear, then, that while the threats to protected areas and rain forest biodiversity are similar all over the world (e.g., habitat destruction, overharvesting, poaching, agricultural encroachment), local conflicts require local solutions. Local community participation is now recognized as essential for conservation and management of biodiversity, yet there are very few examples of areas where such an approach has been successfully applied (see chapter 7). This makes it even more important that the conservation community learn valuable lessons from the solutions that have worked—examples such as the pioneering work of the late Ian Craven in Irian Jaya in Indonesia (Craven and Wardoyo, 1993; Mandosir and Stark, 1993), and Operation CAMPFIRE in Zimbabwe (Mbanefo and de Boerr, 1993). Key lessons seem to be that resolving conflicts requires patience, consultation, and flexibility and that solutions are usually local and linked to the efforts of one or a few crucial individuals.

## LOCAL COMMUNITIES AS STEWARDS FOR CONSERVATION

The Arfak Mountains Nature Reserve in the Bird's Head of Irian Jaya was gazetted as a nature reserve because of its biological importance. It harbors at least 110 species of mammals (53 of them New Guinea endemics) and 320 species of birds, half of the avifauna recorded for Irian Jaya (Craven and de Fretes, 1987). This means that one in 40 of all mammals occurs in the reserve and one in 30 of all birds. Even though the area was gazetted as a strict nature reserve, all the land was traditionally owned by the Hatam people and it was obvious that trying to stop them from collecting resources in the reserve would be counterproductive. Instead, the nongovernmental organization World Wildlife Fund (WWF) worked with local Hatam villagers to develop a management strategy that would enable them to continue their traditional lifestyle but engage them as guardians of the area against outsiders. A WWF team spent more than two years visiting the various tribal villages along the reserve's forested western boundary and patiently negotiating cooperative management agreements with villagers in and around the reserve (Mandosir and Stark, 1993).

The reserve and adjacent outlying lands were divided into 16 nature reserve management areas (NRMAs). The size and boundary of each NRMA were defined by the extent to which each collective group of landowners was willing to work together. A committee of influential people, such as village heads and church leaders, was assigned to manage each NRMA in accordance with tribal customs and community decisions. The committee was responsible for identifying the official landowners and overseeing the correct marking of the boundary. The Hatam were allowed to retain enough land outside the reserve for future subsistence needs. At the same time as the Hatamers were working without pay to mark the western boundary, which all had agreed to, the Indonesian authorities were planting concrete markers along the eastern boundary, without village consultation and following maps that included some village lands and gardens within the reserve. The government's markers have since been removed or ignored, while the Hatam villagers still respect the western boundary.

The management system developed with the Hatam works because the boundary falls under multiple jurisdiction and allows rapid identification of violators as either landowners or outsiders. No one is allowed to establish permanent houses or gardens in the reserve, but the indigenous people are allowed to collect firewood and timber for home use and to hunt with traditional weapons such as bows and arrows. Members of one community may not take forest resources belonging to another community without permission of the owners. Fires may be built for cooking and comfort but not to aid hunting. The regulations allow the continuation of the Hatam's traditional lifestyles, but outsiders face much stricter regulations. They are not allowed to hunt, to make temporary shelters from forest materials, or to remove plants, trees, or animals. Infringements are initially dealt with by the committees, which have government-sanctioned powers to enforce reserve regulations. Violations usually cease after warnings and fines at the community level, but options exist for the committees to pass the matter higher to the reserve management authority (PHPA) or to the district government officer (camat), if necessary. The local communities have taken their responsibilities seriously and have seized the opportunity to play an active role in protecting their own traditional lands and resources.

WWF is also working with the local communities to develop alternative income-generating activities to reduce the need for local people to extend their gardens within the reserve. One such project involves butterfly farming of the famous *Ornithoptera* birdswing butterflies, for which the area is renowned. Gardens of swallowtail food plants, such as the *Aristolochia* vine, have been established in secondary forest areas outside the reserve. Wild butterflies lay their eggs on the vine; larvae feed and pupate high on the vines. The villagers harvest the live pupae, which are then sold to a marketing center. No adult butterflies can be caught or sold. Since only a proportion of the live pupae are found and harvested, wild populations are continually being replenished. With careful control of collection and marketing by the committees and a local nongovernment organization, the butterfly farming should be sustainable. This activity is directly linked to protection of the nature reserve, where wild butterflies spend most of their lives, yet it provides local people with a cash "crop" that is light to transport and yields high returns ($1.50 to $60, according to species) without damaging the natural forest.

The WWF has subsequently begun another project to strengthen local community involvement in park management in Wasur National Park (412,000 hectares in size) in southeastern Irian Jaya. The land within the park boundaries is owned traditionally by 2,000 members of the Kanum, Marind, Marori, and Yei tribes who use it for shifting gardens. Another 65,000 people live around the fringes of the park, many of them subsistence farmers. The need for cash to buy household essentials has led these local communities to engage in small-scale logging, hunting, and selling of land. The WWF project is working with local community groups and government agencies to recognize the park as a traditional-use area where indigenous people and long-term residents are allowed to continue their traditional agricultural and hunting activities.

Part of the management strategy has been to stop illegal hunting with firearms by outsiders while allowing local people to continue hunting traditionally with bows and arrows. Each clan and family has a traditional range where they hunt, garden, and carry out rituals. In 1992, local villagers were able to earn $3,750 in a span of three months from selling deer hunted with bows and arrows. The deer are an introduced species and keeping their numbers in check probably helps native wildlife; the income generated alleviates rural poverty and wins support for protection of the park. Other income-generating options are also being investigated; they include ecotourism and the extraction of essential oils from the bark of the native paperbark (*Melaleuca*) trees. The involvement of local villagers in the protection, management, and controlled exploitation of natural resources is proving an effective tool for conservation (Craven and Wardoyo, 1993).

The Irian Jaya programs have worked well, through a mixture of acknowledgment and extension of traditional rights and customs and the provision of small-scale economic activities. They offer some simple lessons and ingredients for success: close consultation with local people; identification of key players; understanding of community needs; provision of alternative income-generating or social benefits that come "on stream" quickly; strict enforcement of agreed-upon boundaries and regulations, with the communities themselves engaged in enforcement and guarding the reserve; employment opportunities for local people; and flexibility to adapt management strategies to local needs and situations. It might be argued that the situation in Irian Jaya was predisposed to conservation since the Irianese tribes live close to the land and are dependent on natural resources. As community expectations and aspirations change, other management solutions may be needed. It is crucial to ensure that economic incentives or any other benefits are seen to be linked to the park and conservation. Too many supposedly integrated conservation-development projects have turned out to be simply rural development projects with no obvious (or only vague) linkages to conservation (Wells and Brandon, 1992).

## NEW ALLIANCES: INVOLVING INDUSTRY IN PROTECTED AREA MANAGEMENT

Promoting conservation within the wider context of regional development provides an opportunity to broaden its constituency and to establish new partnerships and alliances to help protect and fund parks and protected areas. Local industries are not usually perceived as friends of protected areas, but in certain cases they have much to offer in partnership with conservation agencies: expertise, good resources, and often considerable influence with local and national government agencies. Engaging the private sector in conservation could be of benefit to both business and the environment (MacKinnon et al., 1994).

Kutai National Park in East Kalimantan is one area that could benefit from private-sector interest and investment. Kutai was established in 1936 to protect such species as Sumatran rhino, banteng (*Bos javanicus*), and orangutan (*Pongo pygmaeus*). The reserve was originally 306,000 hectares in size, but large chunks have been excised for logging, oil exploitation, and industrial developments at Bontang and Sangatta. In 1982, Kutai, now reduced to 200,000 hectares, was declared a national park. Nevertheless, illegal logging and agricultural encroachment continued and today much of the eastern part of the park has been cleared by small farmers. Expansion of the oil, natural gas, and fertilizer plants at Bontang led to annexation of more land from the park and drew new immigrants into the area. In spite of this history, Kutai still has considerable conservation value as it is representative of the mixed lowland dipterocarp and ironwood forests once typical of East Kalimantan.

In 1989, a new coal mine was opened at Sangatta and a new road was begun to link Bontang to Sanagatta and Sangkulirang; this road will further open the remaining forests to pioneer farmers. The Sangatta coal mine could have been the "last straw" for Kutai. Instead, it opened new opportunities for the park. Led by Kaltim Primacoal (KPC), owners of the new mine, local industries have agreed to help the park management authority (PHPA) to improve protection and management of the park, in part to safeguard the aquifers and water supplies on which the industries depend (MacKinnon et al., 1994). The park will also provide a valuable recreation area for the industries' employees, including more than 3,000 miners and their families at the Sangatta mine. With sponsorship from KPC, conservation priorities have been identified and development and annual operation plans have been prepared (Petocz et al., 1990). KPC representatives are seeking to draw together a consortium of local industries and international donors to assist PHPA in finding funding for specific management and restoration projects. One potential donor, interested in reforestation of degraded lands within the park, was the Dutch Association of Electricity-producing Companies Foundation, which is funded through a tax on industrial emitters of carbon dioxide in the Netherlands. Large sums of money were promised; a happy ending seemed assured. Then politics intervened. A Dutch minister made one criticism too many, and Indonesia declined all further assistance from the Netherlands and terminated ongoing aid projects. Attempts are still being made to identify other international donors for Kutai. The Kutai initiative was the first example of private industry assisting PHPA with conservation and management, an exciting innovation that is likely to be repeated elsewhere in Indonesia.

## CONSERVING BIODIVERSITY OUTSIDE OF PROTECTED AREAS

While protection of large areas of natural habitats must be the first priority for conservation, much less than 10 percent of tropical forests worldwide are protected within conservation areas. Moreover, in many countries it is unrealistic to hope for more. Not all lands outside protected areas, however, will be converted to agriculture or plantations. There will always be large areas of selectively logged forests, wetlands, fallow lands, and secondary habitats that are valuable for supporting wildlife while also providing important resources to local communities.

Production forests (forests designated for logging) have a key role to play in conservation of biodiversity. By the early 1980s, the area of selectively logged rain forest already exceeded the area of unlogged forest by a ratio of four to one (Brown and Lugo, 1984). This ratio continues to increase. For example, in Indonesia, a country rich in forest resources, half of the total forest area (more than 35 percent of the country) is scheduled to be logged; most of the logged forests will be the species-rich lowland forests that are poorly represented within the protected area network. As primary forest continues to decline, logged forests will have to play an increasingly important role in conservation of many rain forest animals and plants. Selectively logged forests, especially those that are large and only lightly disturbed, can support a high proportion of mature forest species, including most mammals and many species that are unable to survive in isolated forest reserves (Johns, 1988, 1992). The production forests, however, are only useful habitat for wildlife as long as they remain as forests and are actively protected against conversion to other land uses.

Production forests often lie adjacent to existing parks and reserves and can serve as useful buffers and corridors between strictly protected areas, effectively increasing the conservation estate. Production forests can never replace the conservation role of fully protected areas but can be a useful supplement. Since the conservation value of the logged forests will depend on the degree of disturbance to habitats and wildlife during and after logging operations, it is imperative that forest-management practices (including rotation cycles) be improved and revised to minimize disturbance, to maintain biodiversity, and to stop agricultural encroachment after logging. This may require some policy changes, but in many cases stricter application and enforcement of existing regulations would greatly improve forestry practice.

Innovative solutions to conserve biodiversity in production forests are already being tested in Sabah, East Malaysia. The parastatal organization Yayasan Sabah holds a timber concession of almost 1 million hectares. Within this area, two important blocks of lowland and hill forests are designated as conservation areas and will remain intact while the rest of the concession is worked. Danum Valley (43,800 hectares) and Maliau Basin (39,000 hectares) have no official legal status as protected areas but can be regarded as privately established conservation areas, managed by Yayasan Sabah. In cooperation with the Royal Society of the United Kingdom, Yayasan Sabah has established a research station at Danum Valley, where Malaysian and international scientists are engaged in a long-term scientific program (Marsh and Sinun, 1992). Researchers are studying such topics as forest regeneration and the effects of logging on watersheds, forest dynamics, and plant and animal communities (Marshall and Swaine, 1992).

The long-term scientific presence and international collaboration have convinced the Malaysian authorities of the value of the conservation area, so it is likely to be maintained and may even be officially gazetted as a national park. In addition to supporting research, Yayasan Sabah is exploring the possibilities of conducting income-generating activities other than logging within the concession and has established a lodge to serve as a base for wildlife tourism. The success of the tourism venture will depend on maintenance of substantial tracts of lowland rain forests and healthy populations of interesting wildlife, such as orangutan, flying squirrels, and elephants. This in turn should encourage better forest management to minimize disturbance to forests and wildlife during logging operations.

## CHALLENGES FOR THE FUTURE

Conservation of biodiversity has never been so popular, so well publicized, or so well funded. Yet worldwide, an area equal to only about 5 percent of tropical rain forests is devoted to parks and protected areas in which conservation of biodiversity is a primary concern (WCMC, 1992). Forest exploitation will continue and some species losses are inevitable. The problem for the future is how to minimize these losses.

Much of the biodiversity that is saved will lie within protected areas, which in many cases encompass the last remaining blocks of natural rain forests. Effective conservation will require a multiple approach: better protection and management of protected areas; improved environmental management; greater community participation and provision of alternative income-generating opportunities to local communities; increased conservation education and awareness; and, most important, greater political will. Successful conservation will also require better use of existing resources and acceptance of an unpalatable truth: that conservation and utilization are not always compatible. If the global community is sincere in its desire to protect biodiversity and conserve as many species as possible, then some wildlands will have to be fully protected and no exploitation allowed in them. Many countries will need further international support to enable them to implement policies that encourage biodiversity conservation through a spectrum of habitat landscapes, from fully protected to fully utilized.

Even with an expansion of existing protected area networks, much of the world's biological estate will be outside reserve borders, and the fate of many plants and animals—as well as the livelihoods of the people who utilize them—will depend on many factors. These must include efforts to slow deforestation, to log in a more sustainable and less damaging fashion, to protect and manage wetlands for sustainable utilization, and to find alternative and more ecologically sound agricultural options for critical and degraded lands. Scientific research has a key role to play, both in improving our understanding of the linkages among ecosystems and in enhancing crop choice and agricultural production to maximize production on existing farm lands, thus reducing pressure on the remaining protected tropical forests.

It is time for the conservation community to question some existing conservation propaganda and practices. Politicians and international conservation agencies talk glibly of restoring biodiversity, but how feasible is it to restore a tropical forest, and at what cost? Can people and forests live in harmony, where and how? Has there been too much emphasis on the "lose or use" argument for conserving species and on economic justifications of conservation areas? Are extractive reserves and harvesting of nontimber forest products likely to provide long-term economic incentives to conserve tropical forests? What is sustainable utilization and how can it be measured? What lessons, if any, can be learned from traditional agricultural systems, and what are the most suitable crops and cropping regimes for tropical soils? What can be learned from conservation success stories and can these lessons be applied elsewhere? Some of the lessons identified in this chapter for successful protected area management are described in Box 3-1, but much more remains to be learned.

Among the most pressing tasks facing the conservation community is the need to determine whether the large sums of money now available for conservation through international donor agencies are being used effectively. If not, why not? Multilateral development banks and bilateral agencies are keen to fund biodiversity on a hitherto unprecedented scale, but is bigger always better? Lengthy preparations, detailed planning, and multimillion-dollar investments may be appropriate for large infrastructure projects but are less suited for conservation initiatives in which long delays in implementation can lead to disillusionment and loss of interest among affected communities and key stakeholders. Moreover, by the time the project comes to implementation, circumstances may have changed to the extent that the detailed plans may no longer be relevant. In conservation, it is often more efficient to spend less time planning (or even working to existing plans), to be more adaptable during implementation, and to start with small initiatives and build up as different strategies are tried, tested, and modified.

Flexibility is the key to successful conservation management. We can learn valuable lessons about the effectiveness of flexibility from encroachers and other destroyers of biodiversity who adapt rapidly to changing situations, moving their sites and methods of operation to the points of least resistance. The lengthy and rigid negotiation processes and overplanning of large conservation projects, and the slow responses inherent in the bureaucracies of international donor agencies and government departments, all act against flexibility and impede project effectiveness. The great challenge of the next decade will be learning how to mobilize resources better and faster for more effective conservation.

The past 15 years have seen the development and publication of numerous strategies for global, national, and regional actions to conserve biodiversity. Ratification of the Convention on Biological Diversity and Agenda 21 will require countries to prepare more biodiversity strategies and national action plans. Many field conservationists would argue that we have already spent too much time planning, reviewing, and revising, and too few resources on implementing effective management. In many countries, time is running out. The clock is now standing at "one minute to midnight"—the battle to conserve tropical rain forests must be engaged now.

**Box 3-1** Lessons learned: essential ingredients for successful protected area management

Ensuring adequate protection and management of protected areas to conserve biodiversity requires some or all of the following:

- Clear and agreed-upon boundaries of the protected area, with local communities involved in agreement and demarcation (e.g., Arfak Mountains Nature Reserve).
- Clear definition of protected area objectives, and management and zoning according to those objectives.
- Consultation and exchange of information with all stakeholders, especially local indigenous communities, to allow participation in planning and management, building on local knowledge, and creating natural resource management systems as appropriate.
- Firm but fair enforcement of protected area regulations, including regulation through local community structures as well as government agencies.
- Creation of alternative income-generating opportunities to take pressure off the protected area; such activities should be consistent with the conservation objectives of the protected area.
- Identification and engagement of key individuals and community leaders, including women's groups.
- Education and awareness aimed at all levels of stakeholders from local communities to policymakers.
- Institutional strengthening and training for government agencies, nongovernmental organizations, and community groups engaged in protected area management.
- Integration of conservation area in regional development plans to avoid conflicting strategies in different government agencies (e.g., road building).
- Integration of biodiversity conservation in the wider landscape (e.g., adjacent production forests, corridors of natural habitat linking protected areas and production and protection forests).
- Policy reforms to conserve rain forests and biodiversity, and better enforcement of existing policies and regulations.
- Political and financial commitment, including financial mechanisms to cover recurrent funding of protected areas.
- Less planning, faster implementation, and greater flexibility in conservation projects.
- Long-term investment of funds and human resources, with small pilot phase and building on successes and lessons learned.

## REFERENCES

Beehler, B. 1993. *Papua New Guinea: conservation needs assessment,* 2 vols. Port Moresby, Papua New Guinea: Department of Environment and Conservation.

Bibby, C. J., N. J. Collar, M. J. Crosby, M. F. Heath, C. Imboden, T. H. Johnson, A. J. Long, A. J. Stattersfield, and S. J. Thirgood. 1992. *Putting biodiversity on the map: priority areas for global conservation.* Cambridge: ICBP (International Council for Bird Preservation).

Brown, S., and A. E. Lugo. 1984. Biomass of tropical forests: a new estimate based on forest volumes. *Science* 223:1290–93.

Craven, I., and Y. de Fretes. 1987. *Arfak Mountains nature conservation area Irian Jaya management plan 1988–1992.* Bogor, Indonesia: World Wildlife Fund.

Craven, I., and W. Wardoyo. 1993. Gardens in the forest. In *The law of the mother,* E. Kemf (ed.), 23–28. San Francisco: Sierra Club Books.

Davies, G., and J. Payne. 1982. *A faunal survey of Sabah.* Kuala Lumpur: World Wildlife Fund Malaysia.

Diamond, J. M. 1975. The island dilemma: lessons of modern biogeographic studies for the design of nature reserves. *Biological Conservation* 7:129–46.

FAO [U.N. Food and Agriculture Organization]. 1981–1982. *National conservation plan for Indonesia,* 8 vols. Bogor, Indonesia: FAO.

FAO [U.N. Food and Agriculture Organization]. 1990. *Interim report on Forest Resources Assessment 1990 Project.* Rome: FAO.

Hoogerwerf, A. 1949. *De avifauna van de Plantentuin te Buitenzorg.* Buitenzorg: Koninklijke Plantentuin van Indonesie.

Johns, A. D. 1988. Effects of selective timber extraction on rainforest structure and composition and some consequences for frugivores and foliovores. *Biotropica* 20 (1): 31–37.

Johns, A. D. 1992. Vertebrate responses to selective logging: implications for the design of logging systems. In *Tropical rain forest: disturbance and recovery,* A. G. Marshall and M. D. Swaine (eds.), 437–42. London: Royal Society.

Kartawinata, K., H. Soedjito, T. Jessup, A. P. Vayda, and C. J. P. Colfer. 1984. The impacts of development on interactions between people and forest in East Kalimantan: a comparison of two areas of Kenyah Dayak settlement. *The Environmentalist* 4 (Suppl. 7): 87–95.

Lennertz, R., and K. F. Panzer. 1983. *Preliminary assessment of the drought and forest fire damage in Kalimantan Timur.* Samarinda: Transmigration Areas Development Project, German Agency for International Cooperation (GTZ).

MacKinnon, J., and B. Artha. 1982. *A National Conservation Plan for Indonesia,* Vol. VII: *Maluku and Irian Jaya.* Bogor, Indonesia: FAO (U.N. Food and Agriculture Organization).

MacKinnon, J., and K. MacKinnon. 1986a. *Review of the protected areas system in the Indo-Malayan realm.* Gland, Switzerland: IUCN (World Conservation Union).

MacKinnon, J., and K. MacKinnon. 1986b. *Review of the protected areas system in the Afrotropical realm.* Gland, Switzerland: IUCN (World Conservation Union).

MacKinnon, J., and K. MacKinnon. 1987. Conservation status of the primates of the Indochinese subregion. *Primate Conservation* 8:187–95.

MacKinnon, J., K. MacKinnon, G. Child, and T. Thorsell, 1986. *Managing protected areas in the tropics.* Gland, Switzerland: IUCN (World Conservation Union).

MacKinnon, J., and K. Phillipps. 1993. *A field guide to the birds of Borneo, Sumatra, Java and Bali.* Oxford: Oxford University Press.

MacKinnon, J., and J. Wind. 1980. *Birds of Indonesia.* Bogor, Indonesia: FAO (U.N. Food and Agriculture Organization).

MacKinnon, K. S. 1986. The conservation status of nonhuman primates in Indonesia. In *Primates, the road to self-sustaining populations,* K. Benirschke (ed.), 99–126. Berlin: Springer.

MacKinnon, K., G. Hatta, H. Halim, and A. Mangalik. 1996. *Ecology of Kalimantan.* Singapore: Periplus.

MacKinnon, K., A. Irving, and M. A. Bachruddin. 1994. A last chance for Kutai National Park: local industry support for conservation. *Oryx* 23 (3): 191–98.

Malingreau, J. P., G. Stephen, and L. Fellows. 1985. The 1982–83 forest fires of Kalimantan and North Borneo: satellite observations for detection and monitoring. *Ambio* 14 (6): 314–21.

Mandosir, S., and M. Stark. 1993. Butterfly ranching. In *The law of the mother,* E. Kemf (ed.), 114–20. San Francisco: Sierra Club Books.

Margules, C. R., I. D. Cresswell, and O. Nicholls. 1994. A scientific basis for establishing protected areas. In *Systematics and conservation evaluation,* P. L. Forey, C. J. Humphries, and R. I. Vane-Wright (eds.), 327–50. The Systematics Association Special Volume No. 50, Oxford Science Publications. Oxford: Clarendon Press.

Margules, C. R., O. Nicholls, and R. L. Pressey. 1988. Selecting networks of reserves to maximize biological diversity. *Biological Conservation* 50:219–38.

Marsh, C., and W. Sinun. 1992. Pragmatic approaches to habitat conservation within a large timber concession in Sabah. Paper presented at the Fourth World Congress on National Parks and Protected Areas, Caracas, Venezuela, 10–12 February.

Marshall, A. G., and M. D. Swaine. 1992. *Tropical rain forest: disturbance and recovery.* London: Royal Society.

Mbanefo, S., and H. de Boerr. 1993. CAMPFIRE in Zimbabwe. In *The law of the mother,* E. Kemf (ed.), 81–88. San Francisco: Sierra Club Books.

McNeely, J. A. 1988. *Economics and biological diversity: developing and using economic incentives to conserve biological resources.* Gland, Switzerland: IUCN (World Conservation Union).

McNeely, J. A., and K. R. Miller. 1983. *IUCN, national parks and protected areas.* Bangkok: U.N. Economic and Social Commission for Asia and the Pacific.

McNeely, J., and D. Pitt, eds. 1985. *Culture and conservation: the human dimension in environmental planning.* London: Croom Helm.

Medway, Lord. 1971. The importance of Taman Negara in the conservation of mammals. *Malayan Nature Journal.* 24 (2): 212–14.

MoF/FAO [Ministry of Forestry/U.N. Food and Agriculture Organization]. 1991. *Forestry action plan for Indonesia,* 3 vols. Jakarta: Ministry of Forestry.

Nakasathien, S., and B. Stewart-Cox. 1990. *Nomination of the Thung Yai-Huai Kha Khaeng Wildlife Sanctuary to be a UNESCO World Heritage Site.* Bangkok: Wildlife Conservation Division, Royal Forest Department.

Peres, C. A. 1994. Indigenous reserves and nature conservation in Amazonian forests. *Biodiversity Conservation* 8 (2): 586–88.

Peres, C. A., and J. W. Terborgh. 1994. Amazonian nature reserves: an analysis of the defensibility status of existing conservation units and design criteria for the future. *Conservation Biology* 9:34–36.

Petocz, R., N. Wirawan, and K. MacKinnon. 1990. *The Kutai National Park, planning for action*. Bogor, Indonesia: World Wildlife Fund.

Round, P. D. 1988. *Resident forest birds in Thailand: their status and conservation*. ICBP (International Council for Bird Preservation) Monograph No. 2. Cambridge: ICBP.

Stevens, W. E. 1968. *The conservation of wildlife in West Malaysia*. Serembang: Office of the Warden, Federal Game Department, Ministry of Lands and Mines.

Sumardja, E. A., Tarmudji, and J. Wind. 1984. Nature conservation and rice production in the Dumoga area, North Sulawesi, Indonesia. In *National Parks, conservation and development: the role of protected areas in sustaining society*, J. A. McNeely and K. R. Miller (eds.), 224–27. Washington, D.C.: IUCN World Conservation Union/Smithsonian Institution Press.

Van Balen, S., E. T. Margawati, and Sudarayanti. 1988. A checklist of the birds of the Botanical Gardens of Bogor, West Java. *Kukila* 3:82–92.

Van Helvoort, B. E. 1981. *Bird populations in the rural ecosystems of West Java*. Wageningen: Nature Conservation Department.

WCMC [World Conservation Monitoring Centre]. 1992. *Global biodiversity: status of the earth's living resources*. London: Chapman and Hall.

Wells, D. R. 1971. Survival of the Malaysian bird fauna. *Malay Nature Journal* 24:248–56.

Wells, D. R. 1984. The forest avifauna of western Malaysia and its conservation. In *Conservation of tropical birds*, A. W. Diamond and T. E. Lovejoy (eds.), 213–22. Cambridge: ICBP (International Council for Bird Preservation).

Wells, M., and K. Brandon, with L. Hannah. 1992. *People and parks: linking protected area management with local communities*. Washington, D.C.: World Bank, World Wildlife Fund, and U.S. Agency for International Development.

Whitten, A. J., S. J. Damanik, J. Anwar, and H. Nazaruddin. 1987. *The ecology of Sumatra*. Yogyakarta: Gadjah Mada University Press.

World Bank. 1995. Integrating conservation and development: Kerinci-Seblat Integrated Conservation and Development Project. *Facing the Global Environment Challenge* (March–May): 12–13.

# 4

# The Silent Crisis: The State of Rain Forest Nature Preserves

*Carel P. van Schaik, John Terborgh, and Barbara Dugelby*

The principal response of the global community to the threats against biodiversity has been the establishment of strictly protected areas, exemplified by the National Park System of the United States. In such areas, consumptive uses are banned and wild nature is allowed to exist in untrammeled form. Nonconsumptive recreational uses—such as sightseeing, hiking, swimming, boating, and camping—are permitted but are regulated as to place and time and number of participants.

In the tropical forest realm, however, protected nature preserves are in a state of crisis. A number of tropical parks have already been degraded almost beyond redemption; others face severe threats of many kinds with little capacity to resist. The final bulwark erected to shield tropical nature from extinction is collapsing.

The predictable and unpredictable ecological processes likely to affect the future ability of protected rain forest areas to retain their full biodiversity are examined in chapter 3. While the potential impact of ecological processes could be severe, they are amenable to technical solutions and could be solved given sufficient resources and knowledge. A far more immediate and significant threat is posed by human activities.

Indeed, the crisis of parks in the tropics results primarily from increasing human pressure on all unexploited natural resources, aggravated by ineffective protection. Pressure on parks is exerted on local, regional, and national scales, usually taking the form of illegal land appropriation or resource extraction. The attack on tropical parks is being pressed by four main classes of actors: local and displaced populations of agriculturalists and extractors, governments, resource-robbing elites, and (in a few cases) indigenous forest-dwelling populations. In this chapter, we discuss the root causes of the actions of each of these groups and of the institutional failure that results in ineffective enforcement of park legislation.

The inspiration for this chapter came from conversations with fellow tropical field workers. One after another of our colleagues related stories about the areas in which they worked. Their stories contained many common themes, regardless of the geographic area to which they pertained, and one theme in particular stood out: despite legal status and the presence of conservation officers, protected areas are not safe from illegal appropriation and exploitation.

Our perception is that the threat to tropical parks is not widely appreciated. Quantitative information on the loss and deterioration of tropical protected areas is scant, and status reports on individual parks are seldom publicized. Our task, therefore, is to review the meager evidence in order to estimate the magnitude of the threats being faced by protected tropical rain forest areas and to identify the causes of these threats.

Legal definitions of what constitutes a protected area vary from country to country, and most countries have legislated several levels of protection. To enable international comparisons and compilations, the World Conservation Union (IUCN) has drafted a system of categories of protected areas (see MacKinnon et al., 1986). Reflecting our emphasis on the paramount importance of unexploited areas, our focus in the country reviews that follow is on areas that are strictly protected. Areas reserved for nonconsumptive uses fall mainly into IUCN categories I, II, and III; World Heritage Sites; and the core areas of Biosphere Reserves. Limited or appreciable use of biological resources is allowed in the remaining categories. When consulting published statistics on protected areas, it is useful to keep in mind that many of them refer to IUCN categories I through V, which includes areas where some uses are allowed.

## THE STATE OF PROTECTED RAIN FORESTS

### Regional Overviews

Worldwide, almost half of the area originally covered by tropical rain forests has been converted or seriously degraded. The loss of tropical forests has been notably uneven—very little primary forest remains in the Antilles, West Africa, the Atlantic region of Brazil, Madagascar, South Asia (especially Bangladesh), and some Southeast Asian countries (China, Philippines, Vietnam). Trends in tropical forest degradation and loss are being scrutinized by various agencies (see, e.g., Skole and Tucker, 1994). Yet there is little systematic monitoring of the state of the world's tropical parks and nature reserves (but see Machlis and Tichnell, 1985), which collectively make up almost 5 percent of the world's tropical rain forest biome (WCMC, 1992).

Here, we attempt a first effort at providing such an overview. Our purview encompasses only tropical rain forests (evergreen and semievergreen forests of tropical lowlands and montane climates: Whitmore, 1990), and it pertains to the legally defined areas in which no consumptive use of resources is allowed (as described above), which we collectively refer to simply as "parks."

In our surveys, we attempted to gather information on all countries with tropical rain forests. In addition to general information on forest cover, the status of parks, and aspects of management, we present some anecdotal accounts to illustrate the variety of threats to which tropical parks are exposed. In order to generate a quantitative basis for assessment and comparison, we also compiled country reports that follow a standard format. Table 4-1 gives an example of the data compilation procedure for individual countries, and Table 4-2 presents an overview of current threats in all countries for which sufficient reliable information was available.

First, we need to provide a caveat on the quality of the information. A threat had to be deemed "significant" to be reported. In the absence of explicit criteria, different observers undoubtedly varied in their interpretation of the word "significant." In most cases, no independent verification or cross-checking of the information was possible. However, for Indonesia (Table 4-1), two independent sources agreed almost perfectly on the presence or absence of threats but varied somewhat in their assessment of whether threats were significant enough to be included. While this example suggests that variability of the assessment of threats can be expected, it shows that the uncertainty involves the magnitude of threats rather than their existence.

## Central America and Mexico

Although the first conservation areas in this region were established in the 1920s and 1930s, it was not until the past two decades that the process of gazetting parks gathered momentum (Barborak, 1992; IUCN/WCMC, 1992). Because rain forests have largely disappeared in many parts of Central America, only a few additions to the existing system can be anticipated. Lack of active management has led to numerous "paper parks." In several Central American countries, recent civil wars have made effective management and protection impossible (Barborak, 1992). In some countries, park management has been hindered by frequent changes in administrative structure. In Mexico, for instance, more than a dozen agencies have held responsibility for park management during this century (IUCN/WCMC, 1992).

Only about 6 percent of Mexico was originally covered by tropical rain forests; about half of the forest area remains, but little of it is pristine (IUCN/WCMC, 1992). No quantitative assessment of the threats to the parks is available, but at least one, the Volcan Colima National Park, is seriously degraded by hunting, grazing, and logging.

A quantitative assessment of Panama's rain forest parks indicates that poaching and encroachment are common to all of them (G. Adler, pers. comm., 1993). Oil exploration has been permitted in several of them (IUCN/WCMC, 1992). The Cerro Campana National Park exists mainly in name, having been invaded by subsistence farmers.

*Table 4-1* Damage to protected tropical rain forest areas in Indonesia

| Source of damage | | | | | | | | |
|---|---|---|---|---|---|---|---|---|
| Area | Hunting/ fishing | Agricult. encr. | Logging/ fuelwood | Fires | Grazing livestock | Mining | Road building | Hydropower dams |
| Gunung Leuser | Y | Y | Y | N | (Y) | N | Y | N |
| Taitai Batti (Sib.) | Y | ? | ? | N | N | N | N | N |
| Kerinci-Seblat | Y | Y | Y | N | N | Y | Y | P |
| Seberida | Y | Y | Y | ? | ? | N | ? | N |
| Berbak | Y | Y | (Y) | (Y) | N | N | N | N |
| Way Kambas | Y | Y | (Y) | Y | N | N | N | N |
| Barisan Selatan | Y | Y | Y | Y | Y | N | Y | N |
| Ujung Kulon | Y | (Y) | (Y) | N | (Y) | N | N | N |
| Gunung Gede-Pangrango | Y | Y | Y | N | N | N | N | N |
| Meru Betiri | Y | Y | Y | (Y) | N | N | N | N |
| Gunung Palung | (Y) | Y | (Y) | ? | ? | N | ? | ? |
| Danau Sentarum | Y | Y | N | N | N | N | N | N |
| Bukit Baka | N | Y | Y | N | N | N | (Y) | N |
| Tanjung Puting | Y | (Y) | (Y) | N | N | (Y) | N | N |
| Kutai | Y | Y | Y | Y | Y | Y | Y | N |
| Kayan-Mentarang | Y | Y | ? | ? | ? | N | ? | ? |
| Tangkoko | Y | (Y) | Y | Y | N | N | N | N |
| Dumoga-Bone | Y | Y | Y | Y | N | Y | Y | N |
| Lore Lindu | Y | Y | Y | (Y) | N | N | Y | P |
| Rawa Aopa | Y | Y | (Y) | Y | Y | N | Y | N |
| Manusela | Y | Y | Y | N | N | N | Y | N |
| Arfak | (Y) | (Y) | (Y) | (Y) | N | N | N | N |
| Mamberamo | Y | ? | ? | ? | ? | ? | ? | ? |
| Gunung Lorentz | Y | Y | ? | ? | ? | Y | Y | ? |
| Wasur | Y | (Y) | (Y) | Y | N | (Y) | Y | N |
| % Areas affected | 88 | 72 | 48 | 28 | 12 | 16 | 40 | 0 |

*Source:* K. MacKinnon, J. Wind, and C. van Schaik (pers. comm, Dec., 1993)
*Notes:* Y = damage present; N = damage absent; (Y) = damage present but not considered significant; ? = no information (considered N); P = in planning.

***Table 4-2*** Threats to protected tropical rain forest areas

| Country | No. protected areas sampled | Nat'l territory protected (%)[a] | % Protected areas threatened by: Agricult. encr. | Hunting/ fishing | Logging/ fuel wood | Grazing livestock | Mining | Fires | Road building | Hydropower development | Mean no. problems |
|---|---|---|---|---|---|---|---|---|---|---|---|
| Brazil | 30 | 1.71 | 87 | 100 | 60 | 37 | 43 | 30 | 27 | 17 | 4.0 |
| Peru | 6 | 1.97 | 50 | 83 | 50 | 33 | 0 | 17 | 0 | 0 | 2.3 |
| Colombia | 16 | 7.89 | 50 | 63 | 50 | 56 | 13 | 44 | 25 | 19 | 3.2 |
| Ecuador | 6 | 5.62 | 100 | 100 | 100 | 50 | 33 | 33 | 100 | 17 | 5.3 |
| Panama | 9 | 15.22 | 89 | 100 | 78 | 89 | 0 | 22 | 22 | 11 | 4.1 |
| Guatemala | 12 | 7.12 | 42 | 92 | 67 | 42 | 17 | 50 | 50 | 0 | 3.6 |
| Madagascar | 12 | 1.25 | 67 | 75 | 67 | 33 | 0 | 33 | 0 | 8 | 2.8 |
| Zaïre | 5 | 3.64 | 80 | 100 | 40 | 0 | 20 | 20 | 20 | 0 | 2.8 |
| Cote d'Ivoire | 5 | 5.86 | 80 | 100 | 40 | 0 | 80 | 0 | 60 | 0 | 3.6 |
| Rwanda[b] | 3 | 12.42 | 100 | 100 | 100 | 67 | 33 | 0 | 33 | 0 | 4.3 |
| Uganda[b] | 12 | 3.52 | 42 | 100 | 75 | ? | ? | ? | ? | 0 | 2.2 |
| Nigeria[b] | 13 | 2.3 | 77 | 100 | 92 | 15 | 38 | 31 | 23 | 0 | 3.8 |
| India | 12 | 1.4 | 42 | 0 | 25 | ? | 0 | 42 | ? | 33 | 1.4 |
| Sri Lanka | 9 | 7.5 | 78 | 78 | 89 | 89 | 33 | 89 | 11 | 22 | 4.9 |
| Thailand | 21 | 5.53 | 100 | 95 | 48 | 29 | 0 | 0? | 29 | 10 | 3.1 |
| Indonesia | 25 | 7.19 | 72 | 88 | 48 | 12 | 16 | 28 | 40 | 0 | 3.0 |
| Vietnam | 5 | 0.43 | 60 | 100 | 100 | 40 | 0 | 0 | 20 | 0 | 3.2 |

*Sources:* Brazil: Rylands, 1990, 1991; Peru: Terborgh, pers. comm.; Colombia: J. Cavelier, M. Santamaria, and M. E. Chaves, pers. commn., 1994; Ecuador: G. Paz y Miño, pers. comm., 1993; Panama: G. Adler, pers. comm., 1993; Guatemala: Nations et al., 1988; B. Dugelby, pers. comm., 1994; Madagascar: Jenkins, 1987; Zaïre: IUCN, 1987; Sayer et al., 1992; Cote d'Ivoire: IUCN, 1987; Rwanda: D. Watts, pers. comm., 1994; Uganda: Howard, 1991; Nigeria: J. Oates, pers. comm., 1993; India (Karnataka State): Nair, 1991; Sri Lanka: W. Foederer, pers. comm., 1994; Thailand: Rabinowitz, 1993; Indonesia: Table 4-1; Vietnam: Anonymous, 1994.

[a]IUCN categories I, II, and III (WCMC, 1992).

[b]Largely forested reserves (not strictly protected; only illegal logging recorded).

The situation in Guatemala is equally serious. At 1.7 million hectares, the newly declared Maya Biosphere Reserve is one of the region's largest conservation areas. A researcher working there encountered abundant and direct evidence of international collusion to smuggle mahogany out of the reserve. Mexican timber operators pay local Guatemalan authorities handsomely for "rights" to harvest the trees and then transport the logs into Mexico, where they can be freely marketed without arousing suspicion. Oil is currently extracted from Laguna del Tigre National Park, and there are reports that exploitation concessions have been awarded in other parks.

## South America

South America's Amazon basin and Guyana shield contain the largest remaining blocks of rain forest in the world. Protected areas occupy only a small proportion of the landscape, less than 1 percent in the still largely forested Guyanas and variable percentages in the countries forming the Amazon basin, reaching a high of 15 percent in Venezuela. Although much of the land is still covered by old-growth rain forest, the increased influx of disenfranchised highland farmers has recently sent conversion rates soaring in the eastern lowlands of the Andean countries (partly for coca production). Human migration translates into stepped-up pressure on an unprepared protected area system (WRI, 1990).

In several countries, the system of protected areas is still in a nascent state despite a long history of official decrees and legislation. In Bolivia, there is no official list of protected areas, due to a lack of clear definitions, and estimates of the area protected (all habitats) vary from 4 million hectares to 15 million hectares (IUCN/WCMC, 1992). In 60 percent of the delimited parks in Brazilian Amazonia, the land-acquisition process has not been completed, which hampers law enforcement (Peres and Terborgh, 1994). In Ecuador, confusion reigns over the definitions of categories of protected areas; as a result, other government sectors can authorize incompatible activities such as oil exploration. When oil was discovered in the Yasuni National Park, the country's largest, the government promptly signed commercial agreements that allowed a major new production field inside the park. Construction of a road to service the field sparked a major influx of colonists. As recompense, the government has promised that an equivalent area of park land will be created nearby. During the same period, discovery of oil, along with the clearing of forest for oil palm plantations, prompted an official change in the borders of the Cuyabeno wildlife reserve as well (WRI, 1990).

Parks are defended only by token staff in most tropical South American countries. Invariably, park guards are poorly educated, inadequately trained, underpaid, and unarmed and lack the authority to make arrests. In Amazonian Brazil, each park guard is responsible for protecting 6,000 square kilometers, an area larger than the state of Delaware in the United

States (Peres and Terborgh, 1994). Colombia has a mere 200 park guards for its entire system. Some parks in Ecuador, Peru, and Brazil have no guards at all (IUCN/WCMC, 1992).

Lack of protection exposes parks to invaders. Virtually all of Brazil's Amazonian parks are essentially unprotected and are therefore besieged by multiple problems: hunting, fishing, timber poaching, mining, and land occupation. Gold miners are active in one third of all parks and have entirely overrun some parks. When gold miners invaded Brazil's flagship Pico da Neblina National Park, the government responded by sending 2,000 troops to rout them out. Upon entering the park, the expeditionary force found itself outnumbered and outgunned, so it withdrew to a nearby town. The situation developed into a standoff, and the miners continued their activities unimpeded.

In Peru and Colombia, drug traffickers or terrorists have taken over regions containing several parks and have made protection impossible. In general, however, the threats in the Andean countries have yet to reach the same level as in Brazil.

Argentina contains very little tropical rain forest, but the existing parks are relatively intact. Iguazu National Park is subject to some poaching and extraction of hearts of palm, but three other areas are so isolated that few or no problems occur (C. Janson, pers. comm., 1994).

In much of Amazonia, a potential conflict is brewing between the protection of biodiversity and protecting the rights of indigenous people. In Brazil, about one quarter (27 percent) of Amazonian parks contain Indian Reserves, in which the native inhabitants may engage in commercial resource extraction (Peres and Terborgh, 1994). Likewise, in Colombia the status of protected areas and indigenous areas (resguardas) or indigenous reserves is considered compatible (IUCN/WCMC, 1992). Since there are few limitations on the behavior of indigenous inhabitants, whose notion of "traditional practices" is rapidly evolving, there is a risk that unimpeded development will erode biodiversity. Indigenous reserves cover an enormous area in tropical South America. Peres (1994) notes that in the various countries making up Amazonia, almost 100 million hectares are designated as indigenous reserves (exact definitions vary by country), far more than the combined area of all parks. If the legislation covering these areas can be revised to ban large-scale mechanized and commercial resource exploitation, these areas can potentially make an important contribution to conservation through sustainable use.

## Central Africa

Subsaharan Africa contains two major rain forest blocks, one in central Africa and one in West Africa west of the Dahomey (Benin) gap. Establishment of protected areas in the forests of Subsaharan Africa lags behind that of savanna reserves, which are considered greater sources of tourism revenue (Sayer et al., 1992).

In the central forest region, Cameroon retains less than half of its original rain forest. Its most important protected forest, Korup National Park, is severely threatened by pervasive hunting (Sayer et al., 1992). Other Central African countries—Gabon, Equatorial Guinea, Congo, and Zaïre—are still largely forested. The first three of these, however, have set aside only a minute proportion (less than 0.5 percent) of their land area for strict nature preservation. Zaïre has allocated more than 3.5 percent of its territory for preservation, at least on paper. Zaïre also contains more rain forest than any other African country and has the largest conservation areas (WCMC, 1992). Unfortunately, the country is almost without infrastructure and lacks an effective central government, factors that severely impede the management of its five vast tropical forest parks.

Forests around the margins of the Congo Basin are severely threatened. The highland forests of densely populated Malawi, Rwanda, Burundi, and Uganda contain numerous endemic species and are especially vulnerable (WCMC, 1992). In war-torn Rwanda, the most densely populated country in Africa, conservation areas are under severe pressure (D. Watts, pers. comm., 1995). One can only guess at the fate of these areas, as the recent civil war has squelched all law enforcement.

In Uganda, no tropical rain forests were strictly protected until the late 1980s. The highest protection category has until recently been that of forest reserves, in which timber extraction is allowed. Available information (based on Howard, 1991) indicates that all of these areas are subject to hunting and illegal timber extraction, but these assessments are not strictly comparable to those of the other countries because they include forest reserves. A quantitative appraisal of the status of the country's forest reserves showed that only half of the reserved area consisted of undisturbed stands and that all of the reserves were subjected to poaching (Howard, 1991). Even this assessment may be too optimistic (T. Struhsaker, pers. comm., 1995). Fortunately, the Ugandan government recently elevated the status of some forest reserves to national parks and is enforcing this new status.

In Nigeria, little of the original rain forest remains. Apart from a single national park and one wildlife sanctuary in the rain forest zone, most unconverted areas have the status of forest reserves. The situation in these areas is among the worst in Africa (J. Oates, pers. comm., 1994). Although forest reserves should serve for managed timber production and watershed protection, these last tracts of natural forest in the country are rapidly being converted to agricultural uses.

## West Africa

The countries in West Africa have lost virtually all of their rain forests, and their deforestation rates (annual mean of 2.1 percent) are among the highest in the world (Sayer et al., 1992). The amount of rain forest officially preserved is limited to some 2.7 percent of the original area

(Martin, 1991), hardly sufficient to maintain the full biodiversity. In addition to parks, there are forest reserves, but many of these suffer from agricultural encroachment and overhunting, and some have been transformed into timber plantations (Martin, 1991). Thus, at best only a few forest reserves remain to make a serious contribution to future biodiversity protection.

The most important rain forest park in West Africa is Taï National Park in western Côte d'Ivoire (Martin, 1991; Kouadio et al., 1992). Of the nearly 1 million hectares originally gazetted, less than half remains protected, and the remainder is being invaded by "ecological" refugees from the Sahelian zone—nationals displaced by a nearby hydroelectric project, gold mining operations, and timber companies. Together, these groups impose an unsupportable poaching pressure on the park. Other West African parks are subject to multiple threats as well (IUCN, 1987).

In western Ghana, a recent survey failed to locate three primate subspecies endemic to this area and eastern Côte d'Ivoire, whose presence was last recorded in the early 1970s (Struhsaker, 1993). These old-growth specialists are now feared extinct. The forest reserves in which they should occur have been seriously degraded through excessive logging and hunting; much of this pressure is being exerted by displaced people.

In conclusion, the situation in West Africa is among the worst anywhere, with only small patches of rain forest remaining. None of these can be said to enjoy adequate protection.

## Madagascar

Biogeographically, Madagascar is part of the Afrotropical realm, yet it has been isolated from mainland Africa for so long that a distinct flora and absolutely unique fauna have evolved. Sixteen species of primates, several other mammals, and the extraordinary elephant bird disappeared following colonization by humans more than 1,000 years ago (Richard and Dewar, 1991; Goodman, 1994). Most of Madagascar's tropical rain forests have disappeared during the last few decades, with only one third remaining, mostly on steep, inaccessible slopes (Green and Sussman, 1990). Protected areas are still being gazetted, but all are under considerable pressure, mainly from a relentlessly expanding agricultural frontier. Two thirds or more of the protected rain forest areas suffer from encroachment, hunting pressure, and timber extraction—most of the areas from all three threats at once (Jenkins, 1987).

## South Asia

The conservation ethic in this region goes back at least two millennia: in India protected areas have existed since the third century B.C.

(MacKinnon et al., 1986). Nowadays, however, most of South Asia is very densely populated and few areas are protected. Bangladesh has little of its rain forest left, and most conserved forest areas are small and subject to various threats (Collins et al., 1991). Sri Lanka, on the other hand, has declared almost one quarter of its original rain forest estate protected (WCMC, 1992). Many protected areas in Sri Lanka, however, are also threatened. Some of them have been abandoned on security grounds, which has opened them up to significant illicit exploitation (Seneviratne, 1994). Most of the protected areas containing evergreen forest (following Collins et al., 1991) that fit our criteria are under serious threat (W. Foerderer, pers. comm., 1994).

Nair (1991) presents quantitative data for the tropical rain forest national parks and sanctuaries in India's Kerala State. The information suggests that the conservation areas fare somewhat better here than elsewhere in the tropics, especially since religious injunctions serve to inhibit poaching (although the 0 percent recorded is probably overly optimistic; M. Rao, pers. comm., 1992). Elsewhere, mining is a problem. For instance, large chunks of the Sariska tiger reserve in India have been destroyed by large-scale mining operations sanctioned by the government.

Kothari et al. (1989) present the results of a quantitative survey of all of India's parks and sanctuaries. While their figures are consistent with those of Nair, their survey illustrates another remarkable phenomenon. The national park and sanctuary categories confer the highest legal status. Yet, numerous activities occur that seem to go against the *spirit* of the Wild Life [Protection] Act of 1972. For instance, livestock grazing is allowed in 66 percent of the conservation areas, human habitation in 46 percent, agriculture in 43 percent, and timber harvesting in 37 percent. These activities, regardless of whether they are technically legal or not, threaten the biodiversity of India's parks.

## Indochina

Less than one fifth of the original rain forest remains in Vietnam, and less than half in Thailand. Both countries have experienced heavy deforestation recently (Whitmore and Sayer, 1992). The situation in Myanmar, Laos, and Cambodia is somewhat better, as more than half of the forests of these countries remains. Much of this region has witnessed armed conflicts during the past few decades. While Thailand has protected some 18 percent of the area originally covered by rain forest, the other countries in this region have either no formal parks (Laos, Cambodia, Myanmar) or only a few (Vietnam). Conservation plans for these countries are in the formative stage. Reports from Vietnam suggest that very few declared conservation areas currently receive any effective protection and that most, if not all of them, are heavily exploited for wildlife. Some are also threatened by logging, gold mining, large-scale harvesting of medicinal plants, and agricultural encroachment (Eames and Robson, 1993).

In Thailand, illegal exploitation of resources is widespread in parks, and some parks have been further degraded by poorly managed tourism (Rabinowitz, 1993). The country's best-known rain forest park, Khao Yai, is subject to considerable hunting and logging pressure, despite initiatives to curtail these activities (Wells and Brandon, 1992).

## Southeast Asia

The countries of insular Southeast Asia until recently dominated (Philippines) or still dominate (Malaysia, Indonesia) the market for tropical hardwoods and plywood. Most of the region is undergoing an economic boom, accompanied by high rates of deforestation. Both Malaysia and Indonesia still retain approximately two thirds of their rain forests, but less than one quarter remains in the Philippines.

Whereas the Philippines has officially preserved less than 1 percent of its original rain forest area, the figures are better for Malaysia (almost 9 percent), Indonesia (15.6 percent), and Brunei (23.6 percent) (WCMC, 1992). Indonesia, in particular, has embarked on an ambitious plan to preserve the natural environment. The country also has a National Biodiversity Action Plan and is engaging in regionwide land-use planning (see chapter 3). Nonetheless, its protected area system is under no less pressure than are systems in other countries in the region (see Tables 4-1 and 4–2). Government transmigration projects have contributed to the problem by establishing new population concentrations near the borders of parks, thereby aggravating encroachment. Rain forests on the island of Borneo are subject to devastating fires that can destroy millions of hectares, including conservation areas such as Kutai National Park. Fires have increased in frequency and extent because logging operations leave highly combustible slash in the understory and because droughts linked to the El Niño–Southern Oscillation have become ever more intense during the past half century (Salafsky, 1993).

## Australia and Papua New Guinea

Australia and Papua New Guinea (as well as the Indonesian portion, Irian Jaya) still contain most of their tropical rain forests. Australia had only about 1 million hectares of tropical rain forest, and more than two thirds of the forests receive some form of protection (WCMC, 1992). Officially protected areas in Australia are thought to be safe. The situation in Papua New Guinea is radically different, with only 2 percent of the national territory officially protected. It is difficult to make direct comparisons with the situation elsewhere, however, because almost all conservation in Papua New Guinea is in the hands of tribal communities.

## Quantitative Comparisons

The average rain forest park faces between 1.4 and 5.3 significant threats, as illustrated in Table 4-2. The variation is probably based more on variability in reporting than on significant variation in the level of threat among countries. For instance, the low value for India is partly artificial. As explained above, several damaging activities that are illegal in the parks of most countries are legal in India. Thus, the threats to protected rain forest areas are a pantropical phenomenon, not restricted to certain regions.

The stresses reported here are not exaggerated by the inclusion of low-priority areas of little international significance. A recent overview of management problems at World Heritage Sites (WRI, 1992) showed that these premier nature preserves are plagued by the same types problems as the larger set of areas considered here. Although the classification of threats was different, three quarters of the 18 World Heritage Sites in tropical rain forests covered in the report are subject to illegal hunting, almost half to timbering, and some 60 percent to agricultural incursions.

The recorded threats are also not merely a legacy of poor management in the past. Indeed, worldwide, the situation has deteriorated during the past decade. This survey of protected areas—excluding Uganda, Rwanda, and Nigeria, where forest reserves were also included in the survey—suggests that the average protected area is subject to some 3.3 different kinds of threats (based on Table 4-2). If we apply the same categorization of threats to an IUCN survey published in 1984, we find an average of 1.9 threats to tropical parks (mainly, but not entirely, consisting of forests) (IUCN, 1984). Although the comparison must be interpreted with caution because of the different survey methods and the inclusion of non-rain forests in the IUCN sample, the apparent increase during the past decade of almost 75 percent in the number of threats faced per park is ominous. Changes in the intensity or spatial extent of threats are not registered in this survey, but could be increasing even more rapidly.

This overview suggests that most rain forest parks are faltering and are in urgent need of rigorous protection. Extraction of plants and animals is rampant in all regions. Illegal hunting of game for commerce and subsistence is prevalent wherever the killing of animals is not proscribed by religious taboos. And while incursions of agricultural activity into parks may seem a more localized, and hence more limited, problem, numerous potentially destructive activities—such as poaching, livestock grazing, firewood collection, timber extraction, and the setting of fires—typically accompany expansion of the agricultural frontier.

Among the most insidious threats to tropical parks are all-weather roads passing through or near their boundaries. Roads improve access to markets and so create incentives for land conversion and resource extraction. Large commercial and infrastructure projects usually entail road building and so result in the same negative effects. When large projects are concluded, laid-off workers often seek to settle on unoccupied land in the vicinity, thereby adding to pressures on nearby parks.

Do the countries we have examined form a representative sample? Out of necessity, this overview was constructed largely of anecdotal reports. Nevertheless, the survey included the five largest rain forest countries (Brazil, Indonesia, Zaïre, Peru, Colombia) and Conservation International's four megadiversity countries (Brazil, Zaïre, Madagascar, and Indonesia). The absence of reports on the forest parks of additional countries does not indicate a lack of threats and problems, only a lack of available information.

Another possible problem is that the reports that document these patterns are not equally reliable. Is it possible that they exaggerate the seriousness of threats to tropical parks? In general, underreporting is more likely than overreporting. Moreover, the crude form of quantification represented by Table 4-2 may still leave too rosy an impression. The presence of hunting, especially if commercial, is bound to lead to local extinction of at least some of the more vulnerable or sought-after species (Redford, 1992). Fishing is almost ubiquitous, although its effects are less conspicuous than those of hunting. Potentially even more serious is the introduction of exotic fish to the waterways of many nature reserves (Kottelat and Whitten, 1993). Illegal timber extraction has led to the commercial extinction of several tree species, even within some nominally protected areas (Rodan et al., 1992). Thus, wild species harvested for commercial gain are becoming rare or have already gone commercially extinct, even within formal conservation areas—the very areas that were intended to serve as safe havens for such highly sought-after species.

Superficially, some parks may appear intact, with no major logging, hunting, or loss of area. But when one is intimately familiar with the parks, it becomes clear that, even in the most remote corners, people have harvested any accessible commercially valuable resources, sometimes to the point of total depletion (Wind, 1996). Consider, for example, the situation in many parks in Southeast Asia. Commercial-sized rattans of the best variety (*Calamus manan*) can no longer be found. The last surviving rhinos (*Dicerorhinus sumatrensis* and *Rhinoceros sundaicus*) are being slaughtered for their horns. Gaharu (*Aquilaria sp.*) and camphor (*Dryobalanops spp.*) trees are felled for their fragrant heartwood. Certain Lauraceous trees (*Litsea? spec.*) are cut and debarked to provide the raw material for mosquito repellant. Land tortoises, crocodiles, and monitor lizards are harvested for meat and hides. Rivers are overfished with destructive techniques such as poisoning, dynamite, or electrocution, depleting not only fish but also fish-eating mammals, including otters (*Lutra spp.*), flat-headed cats (*Felis planiceps*), and otter-civets (*Cynogale bennettii*). Caves with edible nest–swiftlet (*Collocalia fuciphaga*) nests are sought and exploited. Some highly prized songbirds, especially the straw-headed bulbul (*Pycnonotus zeylanicus*), are captured for the pet trade, a fate also still befalling the occasional orangutan (*Pongo pygmaeus*).

In conclusion, the anecdotes and statistics compiled in Table 4-2 portray the stark reality of the inadequacy of conservation measures in much of the developing world. It will be difficult enough to maintain nature preserves against ecological threats whose sources originate elsewhere (such as introduced species, pollution, increased droughts, or other aspects of climate change). But add to this the mounting pressure of a burgeoning human population in nearly every tropical country, and it is easy to foresee that, unless effective remedial action is taken, many parks will eventually stop serving the objective of preserving biodiversity.

The worst-case examples are several parks that have virtually disappeared, such as Cutervo in Peru and Cerro Campana in Panama. Other major parks have lost one quarter to one half of their originally designated areas; examples include Taï in Côte d'Ivoire, Kerinci-Sebelat and Kutai in Indonesia, and Serrania de Macarena in Colombia. It must be stressed, however, that the latter areas are not necessarily beyond redemption. Park land can be rehabilitated through a combination of rigorous enforcement of park legislation and managed restoration. For example, Shenandoah National Park, located in Virginia in the United States, was mostly worn-out farmland when the park was inaugurated in 1934. Now, a biologically untrained tourist would hardly be aware of the park's human history, were it not for the exhibits prominently displayed in visitor centers. Nevertheless, it is less expensive and biologically more effective to enforce regulations in existing areas than to reconstruct the areas after they have been seriously degraded.

Some have argued that it is too late to replace irreparably damaged parks. Often this is true. The major growth in protected areas was during the 1970s and early 1980s (WCMC, 1992). With most of Earth's land surface now committed to human uses, many possible strategies for preserving biodiversity have already been foreclosed. In some regions, however, opportunities for creating new parks still exist. As shown in Figure 4-1, most tropical countries still have unconverted rain forests that can be added to the pool of protected areas, although the large contiguous blocks of forest that are needed to conserve complete ecosystems are becoming progressively scarcer.

## THE PERPETRATORS OF THREATS: WHO ARE THE ACTORS?

In order to reduce the variety and intensity of threats to rain forest parks, it will be necessary to identify the actors and the motives that drive them. We distinguish four classes of actors that threaten the integrity of tropical nature preserves: local residents, indigenous people, large institutions

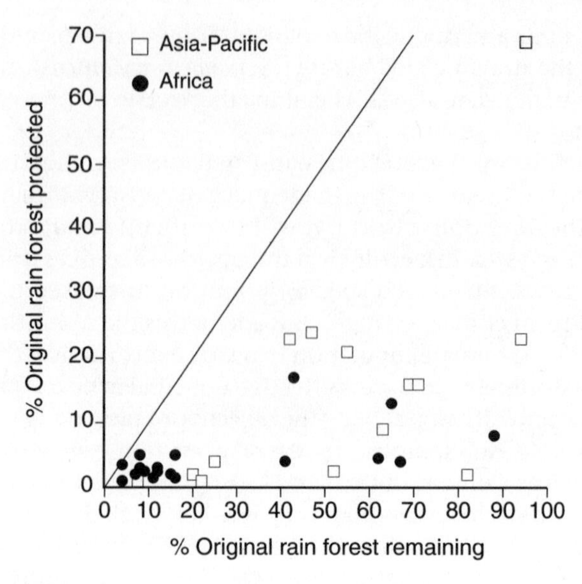


***Figure 4-1*** Remaining versus protected forests: the percentage of a country's original rain forest cover that is protected compared to the percentage of remaining forest. If a country protected all its remaining rain forests, it would be on the diagonal line. Based on Tables 20.9 and 20.10 in WCMC, 1992.

(government and businesses), and resource pirates. In most countries, local people and government-sponsored activities produce the most damage, although the relative roles played by the four sets of actors vary from region to region, and sometimes the distinctions between certain classes of actors become blurred. Here we discuss and offer examples of each category of actor.

## Local Residents

In many regions, parks have become or are becoming natural islands in a sea of human-made habitats. Proximity to human settlements exposes parks to mounting pressure from people eager to make use of the land and its resources. The degradation produced by the activities of people living near parks takes many forms: appropriation of land for swidden (formerly called slash-and-burn) agriculture, hunting, fishing, timber extraction, fuelwood collecting, exploitation of nontimber forest products, mining, livestock grazing, and the setting of fires.

Local populations living near tropical forest parks can be made up of longtime indigenous inhabitants, as in Madagascar, or of recently arrived migrants from other areas, as in the Subsahelian countries of West Africa and in the Andean foothill region of South America. Some of these migrants are ecological migrants—or "shifted cultivators," in the terminology of Norman Myers (1984)—swarming into the only remaining areas of forested land in many countries. Some migrants flee armed conflicts. When violence strikes, rural populations are often displaced from their traditional villages and significantly increase the pressure on unoccupied land elsewhere. This has happened in such countries as Rwanda, Myanmar, Guatemala, El Salvador, and Peru. Still other migrants are brought in as part of government-sponsored resettlement programs, which are widespread in the tropical world, especially where lightly settled tracts of forest still exist. These may bring people into the close proximity of parks, as has happened in Brazil and Indonesia (e.g., Collins et al., 1991).

Some of the people who threaten parks may have land or stable employment elsewhere and are merely looking for sources of additional cash income. These include the so-called truck farmers (Dove, 1993). Other offenders may be people who have been unable to find, or who have lost, regular employment elsewhere in their countries. One such situation has severely stressed the Corcovado National Park in Costa Rica's Osa Peninsula. Workers laid off when a major U.S. fruit company ceased operations in nearby Golfito invaded the park to engage in farming and gold mining. Continuing unemployment of the erstwhile banana workers has required the Costa Rican government to expel illegal miners from the park repeatedly.

The expanding human population and the use of ecologically unsustainable agricultural practices (often required by increased human densities) often leads to permanent forest loss. In Madagascar, for example, the agricultural frontier is advancing rapidly. The highly weathered, phosphorus-deficient soils are unable to recover during short fallow periods, and unproductive wasteland is the end product. Giving up the last few patches of unconverted forest to this process merely delays the moment of ecological collapse.

## Indigenous People

All over the world, rain forests typically harbor indigenous inhabitants whose livelihoods derive from simple agriculture and the extraction of natural products. They are variously referred to as tribal, native, or indigenous people. These people usually live at low densities, use simple technologies, and extract resources only for subsistence. Destructive overexploitation of resources is generally not a part of their cultural heritage.

Increasing exposure of such people to modern technology, however, is rapidly changing this situation. Many traditional forest dwellers are undergoing a rocky transition from a subsistence lifestyle to integration into the market economy. They are also beginning to organize politically and to acquire formal rights to their lands. The combination of political power with land rights may lead to vastly increased pressure on the natural resources of tribal homelands. The Kayapo of

Brazil, for instance, recently acquired title to their lands and promptly began to lease timbering and mining concessions. Timber sales alone brought in $33 million in 1988 (*The Economist*, 1993). Numerous other examples could be cited. The inescapable conclusion is that native peoples are interested in pursuing economic development just as are people without tribal backgrounds. It would be unrealistic and unfair to expect otherwise.

Inevitably, however, the granting of formal rights to indigenous populations living within parks will compromise the integrity of the parks. We noted that many indigenous reserves in Latin America overlap with areas designated for nature conservation. Occasionally, they even coincide, as in the Lacandon Biosphere Reserve in Mexico. The Lacandon Indians were recently given special rights to the land and its natural resources. Having acquired these rights, the Lacandones, like the Kayapo of Brazil, soon began to grant concessions to logging companies.

A less dramatic but no less insidious threat to biodiversity is the rapid growth of indigenous populations. For example, the hill tribe population in the Doi Inthanon National Park in Thailand has more than doubled during the past two decades (Thorsell, 1992). A similar demographic surge is occurring among the indigenous populations of the Manu National Park in Peru. In the long run, population growth will overwhelm any park that contains human populations within its boundaries.

## Large Institutions

National governments, often supported by international agencies or bilateral aid programs, show an affinity for undertaking large projects aimed at improving the economic infrastructure of the country or at developing a whole region. Official indifference to park protection, or a lack of coordination between agencies, can result in the implementation of projects immediately adjacent to parks (Collins et al., 1991). As noted above, conflicts between ministries charged with promoting development and those responsible for protecting parks have all too often been resolved by ignoring or changing the protected status of the land. Mines, hydroelectric dams, timber concessions, transmigration programs, and establishment of plantations are examples.

Roads present particularly notable problems. All large development projects involve road construction, which causes incalculable harm to nature preserves. The U.S. Agency for International Development (USAID) has sponsored many road-building projects. For instance, it financed the paving of a road that bisected the Gunung Leuser National Park on the Indonesian island of Sumatra. Improved access attracted thousands of colonists who settled along the road, severing the link between the park's two main sectors.

### Resource Pirates

Wherever governments fail to enforce conservation legislation, opportunities are created for powerful outsiders—individuals, business corporations, or politically connected cliques—to pillage nature preserves for their resources. Inadequate law enforcement can result from political instability or anarchy, but more commonly results from corruption coupled with a lack of institutional commitment.

Armed conflicts, both within and between countries, are common in the developing world. Needless to say, armies at war seldom respect the boundaries of nature preserves or other legal conventions. In countries burdened by internal armed conflicts—such as Sierra Leone, Angola, Liberia, and Myanmar—one of the warring parties may turn to forest resources to fund their activities. For example, the UNITA (National Union for the Total Independence of Angola) rebels of Angola financed military operations by trafficking in ivory, until the recent international ban on the sale of ivory. Liberian rebels now control the Gola reserve in Sierra Leone and Sapo National Park in Liberia, where they extract timber and hunt game. Several preserves in India, including Manas National Park in Assam, Corbett National Park, and Kanha National Park, are being used by armed rebels as guerilla bases. In Colombia and Peru, several parks have been taken over by drug traffickers.

A lack of strong democratic institutions may allow powerful elites to control access to unexploited natural resources. In countries governed by military regimes, military and civilian elites often collude to extract resources from conservation areas with little regard for legal etiquette. In Zaïre, the army reportedly runs large-scale poaching operations in the country's nature reserves, including the huge Salonga National Park. In many other countries, pressure by business elites may lead to excision of timber concessions or the licensing of extractive uses from nature preserves (Dove, 1993).

Typically, where elites control access to resources, local residents are left to cope with the environmental consequences, such as floods, mud flows, and pollution.

## ROOT CAUSES OF THE CRISIS

We have catalogued the actions that degrade tropical nature preserves, and have identified four primary classes of actors that collectively have put such areas in jeopardy worldwide. Altering the behavior of these groups of actors will require a sophisticated understanding of the incentives that drive them to illegal activity. Here we offer a preliminary analysis of these incentives.

Although many factors are intertwined in motivating the degradation of nature preserves, we can distinguish two major sources of pressure: one that originates within impoverished local populations lacking economic or political power, and one that is generated by the actions of government agencies or powerful, well-connected elites. The distinction is especially important when formulating solutions, because pressures created by individuals can be at least partly addressed at the project level, whereas those generated by governments or elites mainly require action at the (national or international) policy level.

## Pressure from Individuals

As illustrated in Figure 4-2, illegal incursions by local people result from the attraction of unexploited resources (incentives such as those shown in the lower part of the figure) coupled with a lack of resistance (disincentives such as those shown in the upper part of the figure). Individuals make decisions that correspond to the perceived balance of incentives and disincentives. When the former are high and the latter low, violations of park legislation can be expected.

Let us first examine the pressures. In the rural setting of developing countries, children provide support for their parents during old age—a traditional type of social security that operates in the absence of formal pension programs. This leads to large families, thus increasing population density, which in turn inevitably increases the pressure on the land (Dasgupta, 1995). In forested areas where swidden agriculture

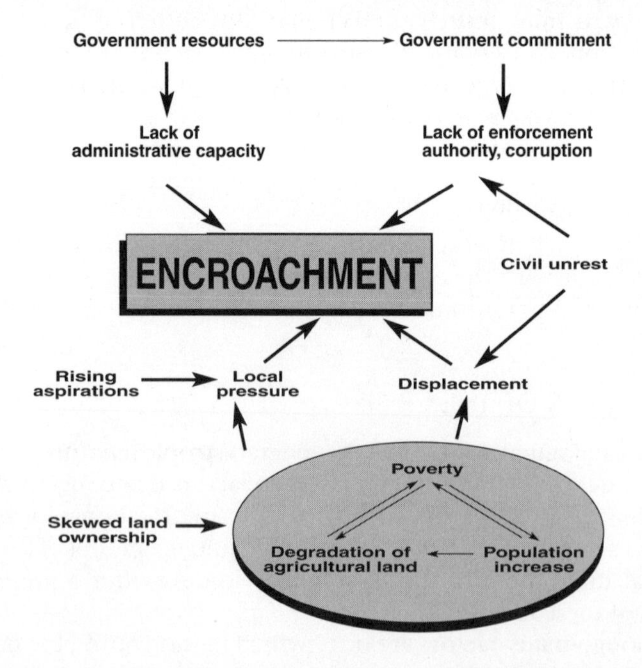


*Figure 4-2* Factors affecting the degradation of protected areas due to actions by local populations, and their mutual relationships.

is the norm, population pressure leads to shortened fallow periods, lower yields, and, eventually, abandonment of the land in exhausted condition. A vicious cycle of poverty, burgeoning populations, and land degradation drives local residents to exploit any unexhausted lands, public or private, that are not physically defended by an owner, or to move on in search of such lands. Inadequately protected nature preserves consequently function as "open access" lands. In Latin America, for example, gross inequities in the distribution of fertile farming land exacerbate the pressure on unused land (Anderson, 1990). Civil unrest may exacerbate this process by providing a steady stream of displaced people.

Finally, as rural communities acquire access to transportation and become exposed to radio and television, their aspirations rise dramatically. The shift from subsistence to market-oriented activity sharply increases the per capita demand on natural resources. While economic development is a fundamental right, it can strain an already overtaxed local production system.

These pressures and processes are familiar and nearly universal among developing countries, yet they would not have such negative impacts on nature preserves that were adequately defended. For instance, as shown in Figure 4-3, there is no simple relationship between the income of the poor and the mean number of threats faced by the forest parks across the countries (taken from Table 4-2). The parallel with deforestation in general is instructive. Deacon (1994) found that one of the important fac-

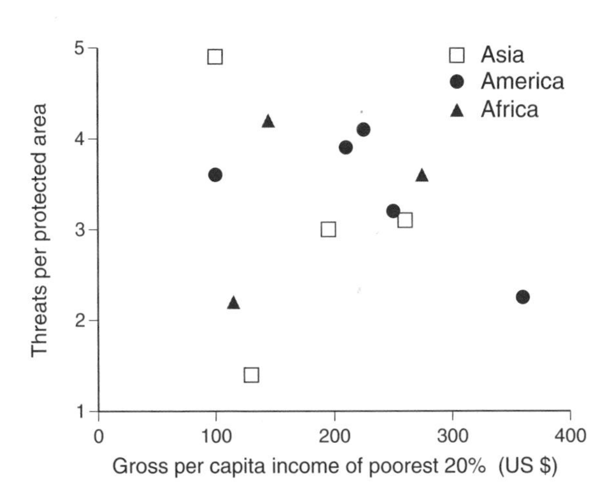


***Figure 4-3*** Threats to protected areas and per capita income: the relationship between the gross per capita income of a country and the degree to which its protected rain forests are threatened (indexed as the mean number of threats per protected area). Threats are taken from Table 4-2; income data from WRI, 1990.

tors affecting deforestation is political instability, which suggests that the presence of well-functioning institutions can act as a brake on deforestation (see also Southgate et al., 1991). Similarly, the relative sanctity of parks in developed countries is not simply a consequence of high incomes. Even in the economically prosperous United States, constant vigilance is required to protect national parks from poachers and gatherers of commercially valuable plants and animals. Yet U.S. parks benefit from staffing levels more than 70 times higher than those of parks in the Brazilian Amazon (Peres and Terborgh, 1994).

Thus, we argue that the defenselessness of tropical forest preserves results at least in part from institutional failure at several levels: parks are inadequately staffed; guards are untrained, unequipped, unarmed, and unable to make arrests; local police forces do not back up park guards by arresting violators; and judges fail to sentence violators. Governments are frequently incapable of defending parks, or unwilling to do so. Where such indifference prevails, lack of financial or administrative capacity and lack of political will are generally interrelated, but in some cases lack of capacity and lack of will operate as independent forces.

A number of rain forest countries have unstable governments and are stressed by varying levels of internal strife. Political instability severely hampers the implementation of a protected area system. The negative effects of political instability are most evident in Indochina, where several countries are only now beginning to set up systems of protected areas. Several Central American and African countries had or have poorly functioning protected area systems for the same reason.

Even where political stability prevails, poverty erodes both the capacity and the will to protect parks. Conservation budgets are meager and guards are unprepared to cope with the cultural and ecological settings within which they are expected to operate. In some countries guards may receive their salaries many months late and are thus obliged to spend most of their time hunting, fishing, and cultivating crops simply to stay alive. With no power of arrest and no backup from police, guards have no incentive to patrol or to investigate suspected violations. Not surprisingly, under such conditions bribery and intimidation are commonplace.

Even in some countries that have established exemplary systems of conservation areas, such as Indonesia and Peru, park guards are limited to filing reports when they discover illegal activities. In many countries, local and national politicians are under pressure from their constituents to downgrade the status of parks. Park managers ignore political pressure from above at their own risk. To keep their jobs, they are often forced to turn a blind eye to violations.

While political expediency or indifference is one cause for lack of governmental commitment, another cause arises out of a chronic tendency to undervalue forests and their biodiversity. Even where poverty is not a major factor, as in fast-developing Thailand and Gabon, there is little evidence of increased commitment to protecting or establishing nature preserves.

## Pressure from Official Projects and Organized Resource Theft

In Figure 4-4, we attempt to analyze the forces underlying large-scale violations of parks. Government-sponsored development projects, typically supported by bilateral loans or multilateral lending agencies, have compromised a number of major rain forest parks. The root causes behind such projects overlap broadly with those that result in a lack of will to defend parks against local encroachment. Government actions are driven by political priorities, and a pressing need to repay foreign debts or support economic development is often paramount. The discovery of oil or a major mineral deposit almost invariably spells the end of protected status. Roads and hydroelectric projects are often planned by government agencies that do not know or do not care about the conservation status of the area affected. Parks departments are easily overruled by politically more powerful agencies.

When governments have been the perpetrators, opposition has often led to public outcry. In contrast, large-scale violation of parks by unscrupulous resource pirates is rarely countered by organized opposition. Yet large-scale resource theft has damaged far more parks than government-

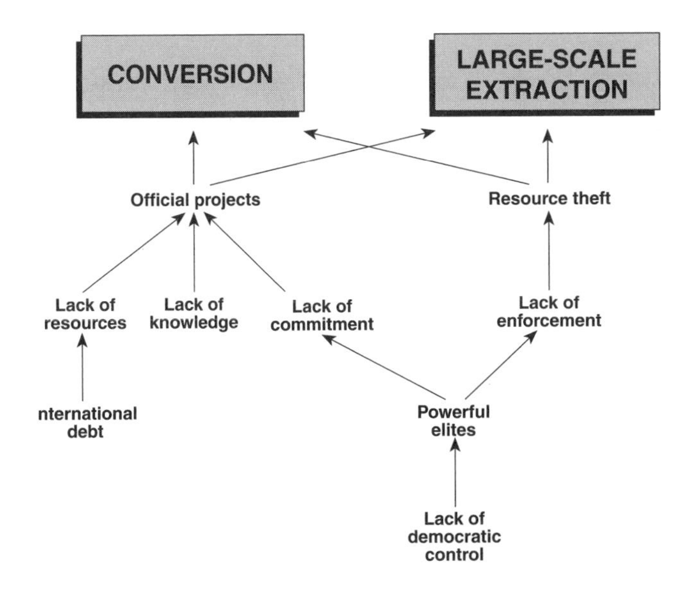


*Figure 4-4* The interacting set of causes underlying the degradation of protected areas due to actions by governments and elites.

sponsored projects, although the two categories of actions cannot always be clearly distinguished. In many countries, collusions of well-connected business people and politicians, sometimes including military commanders, can engage in illegal activities without fear of higher authority. Powerful interests also have pressured some government officials to void the protected status of existing conservation areas, thus enabling the parties to excise timber concessions or proceed with mining or oil extraction inside parks. All too often, the earnings from such large-scale violations do not go to repay international loans or to stimulate the national economy, but end up in foreign bank accounts of the parties responsible for the destruction.

Such abuse of power to gain access to natural resources is common in developing countries. Heavily indebted governments have often stimulated deforestation through subsidies and tax holidays that benefit the well-connected rich rather than the rank-and-file citizen (Gillis and Repetto, 1988). In the worst cases, resource theft occurs without restraint where central power has broken down and where local strongmen or armed bands, eager to exploit any available opportunity to convert natural resources to cash, occupy the countryside. Unfortunately, these situations are bound to become more common in coming years (Kaplan, 1994).

## FORCES PUSHING FORESTS TO THEIR LIMITS

Working separately or together, pressures from individuals and from governments, businesses, and other legal and illegal organizations are pushing rain forest reserves to their limits of sustainability. Biodiversity is suffering accordingly.

Coping with these pressures demands that nations make a coordinated effort to address the complex of root causes, including social and political factors. Only such efforts, begun immediately and pursued with diligence, can stem the swelling tide running against the world's rain forests and the wealth of species that they contain.

## NOTES

We have greatly benefited from discussions with many people, only some of whom we can mention here: Greg Adler, Malcolm Cairns, Jaime Cavelier, Paul Ferraro, Bill Foerderer, Dennis Garrity, Mike Griffiths, Julie Johnson, Randy Kramer, Kathy MacKinnon, John Oates, John Payne, Carlos Peres, Herman Rijksen, Jito Sugardjito, Tom Struhsaker, various members of the van Noordwijk tribe, David Watts, Frances White, and Jan Wind. We thank Madhu Rao for documentation work.

This work was supported by a cooperative agreement between the U.S. Agency for International Development and the Duke University Center for Tropical Conservation. Carel P. van Schaik's fieldwork is supported by the Wildlife Conservation Society.

## REFERENCES


Anderson, A. B., ed. 1990. *Alternatives to deforestation: steps toward sustainable use of the Amazon rain forest*. New York: Columbia University Press.

Anonymous. 1994. Data on protected areas in Vietnam.

Barborak, J. R. 1992. History of protected areas and their management in Central America. In *Changing tropical forests: historical perspectives on today's challenges in Central and South America*, H. K. Steen and R. P. Tucker (eds.), 93–101. Durham, N.C.: Forest History Society.

Collins, N. M., J. A. Sayer, and T. C. Whitmore. 1991. *The conservation atlas of tropical forests: Asia and the Pacific*. London: Macmillan.

Dasgupta, P. S. 1995. Population, poverty, and the local environment. *Scientific American* 272 (2): 40–45.

Deacon, R. T. 1994. Deforestation and the rule of law in a cross-section of countries. *Land Economics* 70:414–30.

Dove, M. R. 1993. A revisionist view of tropical deforestation and development. *Environmental Conservation* 20:17–24.

Eames, J. C., and C. R. Robson. 1993. Threatened primates in southern Vietnam. *Oryx* 27:146–54.

*The Economist*. 1993. The savage can also be ignoble. 12 June: 12:54.

Gillis, M., and R. Repetto, eds. 1988. *Public policy and the misuse of forest resources*. New York: Cambridge University Press.

Goodman, S. M. 1994. The enigma of anti-predator behavior in lemurs: evidence of a large extinct eagle on Madagascar. *International Journal of Primatology* 15:129–34.

Green, G. M., and R. W. Sussman. 1990. Deforestation history of the eastern rain forests of Madagascar from satellite images. *Science* 248:212–15.

Howard, P. 1991. *Nature conservation in Uganda's tropical forest reserves*. IUCN [World Conservation Union] Tropical Forest Programme. Gland, Switzerland: IUCN.

IUCN [World Conservation Union]. 1984. Categories, objectives and criteria for protected areas. In *National parks, conservation, and development: the role of protected areas in sustaining society*, J. A. McNeely and K. R. Miller (eds.), 47–53. Washington, D.C.: Smithsonian Institution Press.

IUCN [World Conservation Union]. 1987. *Directory of Afrotropical protected areas*. Gland, Switzerland: IUCN.

IUCN/WCMC [World Conservation Union/World Conservation Monitoring Centre]. 1992. *Protected areas of the world: a review of national systems*, Vol. 4: *Neoarctic and neotropical*. Cambridge: IUCN.

Jenkins, M. D. 1987. *Madagascar—an environmental profile*. Gland, Switzerland: IUCN.

Kaplan, R. 1994. The coming anarchy. *Atlantic Monthly* (Feb.): 44–76.

Kothari, A., P. Pande, S. Singh, and D. Variava. 1989. *Management of national parks and sanctuaries in India: a status report*. New Delhi: Indian Institute of Public Administration.

Kottelat, M., and A. J. Whitten, with Sri Nurani Kartikasari and Soetikno Wirjoatmodjo. 1993. *Freshwater fishes of Western Indonesia and Sulawesi*. Singapore: Periplus.

Kouadio, E. B., D. F. N'Goran, and F. Lauginie. 1992. La sauvegarde du parc national de Taï: une responsabilité internationale. In *Conservation of West and Central African rainforests*, K. Cleaver, M. Munasinghe, M. Dyson, N. Egli, A. Peuker, and F. Wencélius (eds.), 169–76. Washington, D.C.: World Bank.

Machlis, G. E., and D. L. Tichnell. 1985. *The state of the world's parks: an international assessment for resource management, policy, and research*. Boulder, Colo.: Westview Press.

MacKinnon, J., K. MacKinnon, G. Child, and J. Thorsell. 1986. *Managing protected areas in the tropics*. Gland, Switzerland: (World Conservation Union) IUCN.

Martin, C. 1991. *The rainforests of West Africa: ecology, threats, conservation*. Basel: Birkhäuser.

Myers, N. 1984. *The primary source: tropical forests and our future*. New York: Norton.

Nair, S. C. 1991. *The southern Western Ghats: a biodiversity conservation plan*. New Delhi: Indian National Trust for Art and Cultural Heritage.

Nations, J. D., B. Houseal, I. Ponciano, S. Billy, J. C. Godoy, F. Castro, G. Miller, D. Rose, M. R. Rosa, C. Azurdia. 1988. *Biodiversity in Guatemala: Biological diversity and tropical forests assessment*. Washington, D.C.: World Resources Institute.

Peres, C. A. 1994. Indigenous reserves and nature conservation in Amazonian forests. *Conservation Biology* 8:586–88.

Peres, C. A., and J. W. Terborgh. 1994. Amazonian nature reserves: an analysis of the defensibility status of existing conservation units and design criteria for the future. *Conservation Biology* 9:34–36.

Rabinowitz, A. 1993. Estimating the Indochinese tiger, *Panthera tigris corbetti*, population in Thailand. *Biological Conservation* 65:213–17.

Redford, K. H. 1992. The empty forest. *BioScience* 42:412–22.

Richard, A. F., and R. E. Dewar. 1991. Lemur ecology. *Annual Review of Ecology and Systematics* 22:145–75.

Rodan, B. D., A. C. Newton, and A. Verissimo. 1992. Mahogany conservation: status and policy initiatives. *Environmental Conservation* 19:331–38.

Rylands, A. B. 1990. *Evaluation of the current status of federal conservation areas in the tropical rain forest of the Brazilian Amazon*. Unpublished report to the World Wildlife Fund. Washington, D.C.

Rylands, A. B. 1991. *The status of conservation areas in the Brazilian Amazon*. Washington, D.C.: World Wildlife Fund.

Salafsky, N. N. 1993. The forest garden project: an ecological and economic study of a locally developed land-use system in West Kalimantan, Indonesia. Ph.D. diss., Duke University.

Sayer, J. A., C. S. Harcourt, and N. M. Collins. 1992. *The conservation atlas of tropical forests, Africa*. New York: Simon and Schuster.

Seneviratne, D. 1994. Management crisis in Sri Lanka's wildlife reserves. *Counterpoint* (Mar.): 6–14.

Skole, D., and C. Tucker. 1994. Tropical deforestation and habitat fragmentation in the Amazon: satellite data from 1978 to 1988. *Science* 260:1905–10.

Southgate, D., R. Sierra, and L. Brown. 1991. The causes of tropical deforestation in Ecuador: a statistical analysis. *World Development* 19:1145–51.

Struhsaker, T. T. 1993. *Ghana's forests and primates. Report of a field trip to Bia and Kakum National Parks and Boabeng-Fiema monkey sanctuary in November 1993*. Report to Conservation International, Washington, D.C.

Thorsell, J. 1992. *World heritage twenty years later.* Gland, Switzerland: IUCN (World Conservation Union).

WCMC [World Conservation Monitoring Centre]. 1992. *Global biodiversity: status of earth's living resources.* London: Chapman and Hall.

Wells, M., and K. Brandon, with L. Hannah. 1992. *People and parks: linking protected area management with local communities.* Washington, D.C.: World Bank, World Wildlife Fund, and the U.S. Agency for International Development.

Whitmore, T. C. 1990. *An introduction to tropical rain forests.* Oxford: Clarendon Press.

Whitmore, T. C., and J. Sayer. 1992. Deforestation and species extinction in tropical moist forests. In *Tropical deforestation and species extinction,* T. C. Whitmore and J. Sayer (eds.), 1–14. London: Chapman and Hall.

Wind, J. 1996. The Gunung Leuser National Park: history, threats, and options. In *The Gunung Leuser National Park, Sumatran Sanctuary,* C. van Schaik and J. Supriatna (eds.). Jakarta: The Indonesian Foundation for the Advancement of Biological Sciences.

WRI [World Resources Institute]. 1990. *World Resources 1990–91.* New York: Oxford University Press.

WRI [World Resources Institute]. 1992. *World Resources 1992–93.* New York: Oxford University Press.

# 5

# Policy and Practical Considerations in Land-Use Strategies for Biodiversity Conservation

*Katrina Brandon*

Much attention has been given to the issues of sustainable use, sustainable development, and biodiversity conservation, as well as to the relationships among them. Some observers express a sense of optimism that implementing sustainable activities worldwide will lead to the conservation of biodiversity. In the popular media, there are examples almost daily of conservation success stories. But publicity for conservation and attention on biodiversity are being mistaken for solutions.

What is perhaps more sobering than equating publicity with actions, or actions with solutions, is that the entire rubric of sustainability, in the rural context, has a set of questionable assumptions that underlies the portfolio of activities being implemented to conserve biodiversity. These assumptions have had a major impact in shaping the range of activities that have been developed to address the conservation of biodiversity—from policies (such as the Biodiversity Convention adopted at the 1992 Earth Summit) to projects implemented by conservation and development organizations worldwide (such as the Global Environmental Facility). The questionable assumptions fit into the following seven broad categories:

- *Method.* Biodiversity conservation can best be accomplished through field-based activities, such as establishing parks and reserves.
- *Use.* Sustainable use is possible under a variety of management regimes ranging from private to communal. Dependence on wildlands resources is most likely to ensure their long-term conservation.
- *Incentives.* Appropriate sets of incentives can be readily defined and will influence people to conserve biodiversity.
- *Management.* Management should be devolved to local control whenever possible.
- *Technology.* Technical and organizational solutions exist to improve resource management and production activities in areas with great biodiversity.

- *Poverty Mitigation and Development.* Rural poverty-mitigation and development strategies will lead to conservation and maintain biodiversity.
- *Social.* Local people are cooperative and live in harmony with one another and with nature.

These assumptions, and their implications, are increasingly being questioned as concern mounts that their implementation may lead to serious loss of biodiversity (see Ludwig et al., 1993; Robinson, 1993). While each of the assumptions may be true in certain cases, they have collectively led to the implementation of many conservation efforts that, while attempting to address both human needs and conservation policy, are unlikely to show substantive gains in either.

This chapter explores these assumptions and what they have meant for current thinking about biodiversity policy. Particular emphasis is given to strategies for protecting parks and other protected areas. The chapter concludes with a discussion of what initiatives should be supported were these assumptions to be eliminated from the center stage of biodiversity conservation.

## THE LANDSCAPE OF BIODIVERSITY CONSERVATION EFFORTS

Efforts to preserve biodiversity are being promoted in a wide array of land-use projects. Parks and protected areas (collectively referred to here as parks) are viewed as only one small response to biodiversity conservation. Parks will only protect a fraction of the world's species and ecosystems. And while parks must be made to "work well," it is clear that we need to examine the range of alternatives for managing areas adjacent to or near parks, which also may be subject to some restrictions to encourage biodiversity conservation. For areas adjacent to parks, two types of very different management categories can be broadly defined: integrated conservation-development projects (ICDPs) and locally managed reserves. ICDPs are areas or projects in which the defining factor is some linkage between what happens in a core area (the protected zone) and in adjacent areas (Wells and Brandon, 1992). These land uses include biosphere reserves, forestry reserves, multiple-use areas, bufferzone projects, and large-scale planning units such as regional conservation areas.

Locally managed reserves do not have sustainable use as their primary objective. In most situations, human (consumption) objectives come first. Reserves generally rest on the assumption that ecological, social, and economic needs are on par with one another—but in the final analysis, the groups managing the reserves have a say over how management will proceed. Examples of locally managed reserves include indigenous reserves and extractive reserves. In one sense, the term "reserve" gives a hint of their different function: reserves seek to "set aside" areas for tribal groups or for extraction of resources. In contrast, ICDPs promote the "integration" between parks and adjacent lands.

Each of these conservation units and their primary objectives are summarized in Table 5-1.

## Parks

Worldwide, parks represent the most important method of conserving biological diversity in situ. Recognition of their role in the protection of species, habitats, and ecosystems has led to a dramatic expansion of parks during recent decades, with more parks and reserves having been established since 1970 than in all previous periods. By 1989, about 4,500 sites—covering 4.8 million square kilometers (3.2 percent of Earth's land surface)—had been designated as protected areas. About 2,250 of these sites, covering 2.4 million square kilometers, are located in the tropics; national parks comprise 792 of these sites and occupy 1.3 million square kilometers (Reid and Miller, 1989). Although the World Conservation Union (IUCN) lists eight categories of parks and protected areas, few of the categories have biodiversity conservation as their primary concern. IUCN categories I, II, and IV do have biodiversity conservation as a primary objective; however, even categories such as II, which includes national parks, may be zoned for all kinds of uses, depending on national policies. Many of the other categories preserve lands for historic and cultural values, recreational use, and tourism.

Many parks, particularly those in the tropics, are experiencing serious and increasing degradation as a result of large-scale development

*Table 5-1* Summary of conservation units

| Unit | Objective |
|---|---|
| Parks | Biological criteria come first. Wildlands and wildlife protection is the key objective (IUCN park categories I, II, IV). |
| ICDPs | Parks and the areas surrounding them are linked through an array on incentives, conservation and development activities, and management strategies. New management authorities are often created. Use is promoted subject to external controls. |
| Locally managed reserves | Resident people can preserve biodiversity through activities under their control and largely subject to their management regimes. Examples include both indigenous and extractive reserves. |

projects, expanding agricultural frontiers, illegal hunting and logging, fuelwood collection, and uncontrolled burning. Many of the underlying reasons for the alarmingly high rate of loss of biodiversity are ultimately attributable to policies in a variety of different sectors. While local people are often regarded as "the problem" or the proximate source of threat, their actions are often influenced by factors over which they have little control. In some areas, the result is that parks are increasingly becoming natural islands surrounded by a variety of land uses, including colonization, agriculture, plantation forestry, and cattle grazing.

## Integrated Conservation-Development Projects

ICDPs link the conservation of biological diversity in parks with local social and economic development (Wells and Brandon, 1992). They aim to achieve conservation by promoting socioeconomic development and by providing local people with alternative sources of income that do not threaten to deplete the flora and fauna of the parks. The various approaches described under the rubric of ICDPs (see Box 5-1) are based on concepts of sustainable use and sustainable development in the rural context. They imply types of land-use alternatives that, in combination with a range of social, technical, and economic options, will lead to conservation of biodiversity. Implicit in these approaches is an assumption that biodiversity conservation may not be as great in ICDPs as within strictly protected areas; nevertheless, the argument is that allowing the use of resources in these areas places a value on the resources, and this ascribed value will help lead to conservation of the resources.

The view that underlies ICDPs was underscored at the 1982 World Parks Congress in Bali—that "protected areas in developing countries will survive only insofar as they address human concerns" (Western and Pearl, 1989:134). In all cases, ICDPs seek to define and enforce the appropriate management levels that will lead to sustainable use. ICDPs try to link uses outside parks with what happens inside the parks.

All ICDPs have a "core" protected area in which uses are restricted. The implicit assumption is that these core areas will remain intact only if people are restricted to using the outlying areas. There are three major strategies that ICDPs have attempted: enhancing park management and/or creating buffer zones around protected areas; providing compensation or substitution to local people for lost access to resources; and encouraging local social and economic development.

## Locally Managed Reserves

Locally managed reserves are areas in which the emphasis is on maintaining livelihoods; biodiversity conservation is a secondary objective. Both extractive and indigenous reserves are large tracts of land, often

**Box 5-1** Examples of integrated conservation-development projects

Integrated conservation-development projects (ICDPs) are land uses, activities, and projects whose primary aim is to foster conservation in parks by promoting socioeconomic development to provide local people with livelihoods that do not threaten the parks' resources. They are based on the premise that management of protected areas must reach beyond traditional conservation activities inside park and reserve boundaries to address the needs of local communities outside the boundaries (Wells and Brandon, 1992). Some examples are given below.

- *Biosphere reserves* first appeared in 1979. They have a protected "core area" surrounded by a "buffer zone" and then a "transition area." Buffer-zone uses are limited to activities compatible with the core area, while development activities involving local communities are intended to take place in the transition area (Batisse, 1986). There are now approximately 275 biosphere reserves worldwide.
- *Multiple-use areas* are intended to sustain the production of water, timber, wildlife, pasture, and outdoor recreation. The conservation of nature is oriented to supporting economic activities. In some cases, zones are delineated to achieve specific conservation objectives; when this occurs, multiple-use areas are virtually identical to biosphere reserves, except that they lack the official designation. The Annapurna Conservation Area in Nepal is such a multiple-use area.
- *Buffer zones* are areas "peripheral to a national park or equivalent reserve, where restrictions are placed upon resource use or special development measures are undertaken to enhance the conservation value of the area" (Sayer, 1991:2). Activities envisioned for buffer zones usually include hunting or fishing using traditional methods, collecting fallen timber, harvesting fruit, seasonal grazing of domestic stock, and cutting bamboo, rattan, or grasses. In ecological terms, buffer zones promote land uses and practices that are compatible with contiguous parks.
- *Large-scale planning units*, such as regional conservation areas, attempt to link core protected areas, such as parks, with a progression of land uses. Uses compatible with biodiversity conservation, such as managed forests, are located next to parks, while less compatible activities (roads, dams, urban land uses) are located farther away. Examples include the Regional Conservation Area System in Costa Rica (see Umaña and Brandon, 1992) and CAMPFIRE in Zimbabwe.

in areas of high biodiversity, where local people have been granted rights to live and to use forest products. Historically, governments have set aside locally managed reserves—as well as parks—because the lands were considered too remote or too marginal to be worth maintaining.

Indigenous reserves are primarily large tracts of land that governments have turned over to the groups who have lived there for centuries. In Colombia, for example, 26 million hectares have been set aside for indigenous communities (Sanchez et al., 1982). Extractive reserves are usually large tracts of land where local populations "extract" an array of products for both subsistence and sale. Most of the experience with extractive reserves is based on examples from Brazil, where rubber-latex, Brazil nuts and cashews, fruits, and other nontimber forest products are extracted by local residents and sold in regional, national, and international markets. A recent examination of Brazilian extractive reserves notes that "there was as much preoccupation with social issues as there was with the maintenance of biodiversity [and that] social conditions are guiding the decisions about when and where to create extractive reserves" (Martins, cited in Seymour, 1994).

Governments limit the uses permitted in extractive reserves; few such restrictions are placed on indigenous reserves. Indigenous groups generally have the authority to make decisions about the land and resources, although in some countries environmental conditions may be a prerequisite of land transfer. Also, governments have not respected these decisions when they have stood in the way of government development plans such as petroleum exploration or road construction. While people living on extractive reserves have limits imposed externally (e.g., on logging), within those limits they largely control the type and level of use.

## Linking Rhetoric to the Biodiversity Landscape

In practical terms, we would like to think that we are knowledgeable about how to plan and execute conservation activities in these three types of units. But there has been a tremendous lack of conceptual clarity about the categories themselves, the objectives in each category, and what the lack of clarity means for implementation strategies. The resultant muddled projects are not the result of malice or sloth. Rather, they result from a sense of optimism that we can have it all, that is, biodiversity conservation and development. The danger is that these positions continue to be reiterated on a national and international basis, as reflected in such policies and planning documents as *Caring for the Earth* (IUCN et al., 1991), the 1992 World Parks Congress, and the Brundtland Report (WCED, 1987).

But as increasing numbers of evaluations of these activities suggest, we do not know all that we think we do, and many of today's field-based initiatives are not living up to their proclaimed potential. (For more than 60 case studies, see West and Brechin, 1991; Wells and Brandon, 1992; Western and Wright, 1994.) These shortcomings are largely due to a belief among conservationists that what they are doing is conservation—when, in fact, they are really doing large-scale social interventions in complicated settings. Only lately has there been acknowledgment that each of three types of conservation units has distinct design and implementation elements. The following section discusses the questionable assumptions previously identified and the effects they have had on biodiversity conservation.

## THE ECOLOGICAL FALLACY APPLIED TO BIODIVERSITY

Most graduate-level courses in research methods teach students about the "ecological fallacy." The ecological fallacy occurs when a researcher (or someone else) applies the results of a study to a different population than the study analyzed. In the social sciences, for example, it would be invalid to study food preferences of a population of white males over the age of 50 and declare that their preferences extend to white males of any age, or to all men, or to women.

Many of the assumptions put forth at the beginning of this chapter are, unfortunately, ecological fallacies—applied to ecology and conservation. Part of this ecological-ecological fallacy may result from a backlash against how conservation was being practiced several decades ago. Many of these assumptions were first linked to conservation in the World Conservation Strategy (IUCN et al., 1980) and were highlighted at the Bali World Parks Congress in 1982. At that time, a great deal of development theory and rhetoric was focused on dependency theory, the legacy of colonialism, and imperialism. There was a backlash against applying concepts, such as national parks, from developed countries and importing them wholesale onto the landscapes and cultures of the developing nations. The role of policies, the importance of considering the needs of local peoples living in or adjacent to parks, the perspectives of "appropriate and small-scale" technologies, local empowerment, popular participation, democratization and devolution of power, and an emphasis on harmony—all important considerations—fit well with the need to internalize development concerns into the conservation debate.

Unfortunately, these issues were often imported without thought to their implications on a site-specific basis. Over time, these assumptions have become rooted in nearly all of the field-based conservation activities being tried—whether they be parks, ICDPs, or locally managed reserves. We are now realizing that many of the assumptions, despite their good intentions, are flawed. More evidence is showing that most of the assumptions do not hold most of the time. And the fact that we are actively designing conservation agendas around these assumptions is likely to lead to some significant failures in biodiversity conservation. Examples of how little we know, and where these flawed assumptions have backfired, are given below.

### Method

The assumption is that biodiversity conservation can best be accomplished through field-based activities, such as establishing parks and reserves.

A wide variety of policies affect biodiversity conservation in developing countries. Most frequently cited are pricing policies for agricultural commodities, fuels, and wood; transportation and national integration policies such as road building; subsidies for cattle ranching and other land uses that may not be appropriate; and land-tenure policies that encourage colonists to settle frontier areas while allowing productive lands to remain underutilized. Furthermore, most reviews of case studies and field-based projects suggest that in the absence of broader political will, specific projects are unlikely to achieve their objectives. According to Seymour (1994), "Community (conservation) initiatives are strongly influenced by the constraints and opportunities created by the macro policy environment . . . projects start out focused on specific sites, but sooner or later have to deal with structural and policy issues" (pp. 494, 495).

This is important because more project-level interventions are being targeted in countries than ever before (see Southgate and Clark, 1993). Yet there has been little attention to identifying what the "right" policies are for different countries and how to implement them. One clear example of the failure to promote strong policy reform as a serious initiative comes from the World Bank. The Bank is significant because it is the largest environmental agency and lender in the world and because of its influence on other organizations and governments. As of the 1993 fiscal year, the Bank was financing nearly 100 environmental projects in more than 40 countries, representing commitments of nearly $5 billion. The Bank's World Development Report, *Development and the environment* (1992), emphasized the need to focus on "win-win" policies—policies that make good economic sense, generate additional revenue for countries, and are environmentally friendly. The magnitude of promoting "win-win" policies can be substantial; for example, between 1975 and 1986, Brazil had a hidden subsidy of more than $1 billion for livestock ranching, the "biggest known subsidy in history for ecological destruction, unrelieved by economic gain" (*The Economist*, 1989). Removing such a subsidy represents significant "wins" for the economy and the environment. Yet in its policy dialogue with borrower countries, the Bank has, in general, been timid in identifying these policies or in pushing for their reform (Brandon, 1995).

Similarly, there has been no concerted push by nongovernmental organizations (NGOs) for the World Bank to identify what types of interventions are most likely to support rural-focused and "green" policies—that is, those policies likely to have the greatest impact on biodiversity. In what is perhaps an ironic twist, the environmental NGOs, in testimony before the U.S. Congress, have provided specific recommendations on how to increase the Bank's attention to poverty alleviation as part of the lending process and on how to focus adjustment policies to directly benefit the poor. Often, it seems as though the environmental NGOs are more concerned with social issues than with environmental issues. And when environmental issues are raised by the NGOs, the U.S. Congress, or the U.S. Department of Treasury, the issues typically represent only a small part of those that could be raised. The attention to sustainable development and poverty alleviation, while important, does not directly address environmental issues—much less biodiversity conservation, in particular. If anything, these concerns broaden the agenda and dilute the message.

Field-based projects are important for conservation—in many countries, they are the only thing providing protection for or management of protected areas. In essence, they may be literally "holding ground" while policy issues are addressed. Nevertheless, there are three problems that arise from carrying out projects without creating effective policies. First, projects often cannot achieve their objectives unless the policy context is "correct." Second, there is little concerted effort to define and aggressively push appropriate biodiversity-conservation policies at the country level. Third, the environmental groups promoting attention to policy are not stressing attention to environmental policy, much less to biodiversity conservation. The lesson is that if we do not pay attention to both policy and field-based initiatives, we are unlikely to achieve success.

## Use

The assumption is that sustainable use is possible under a variety of management regimes ranging from private to communal. Dependence on wildlands resources is most likely to ensure their long-term conservation.

The notion of sustainable use—and the importance of making protected areas relevant to local communities—got its first major push in the 1980 World Conservation Strategy. The concern was that parks were under considerable threat because they did not serve the needs of local communities, which reflected a conviction that conservation could best be accomplished if people could wisely use and exploit resources. The most important formulation of the need for sustainable use is contained in *Caring for the Earth: A Strategy for Sustainable Living*, prepared by the World Conservation Union, the United Nations Environment Programme, and the World Wildlife Fund (IUCN et al., 1991). In proposing this strategy, the international conservation organizations have gone a long way toward promoting the myth that development and the conservation of biodiversity are compatible.

Regarding parks, proponents of sustainable use argue the need to link "protected areas together with human needs [to] support ecologically sound development which takes on practical meaning for governments and local people . . . part of a vision of the future where we can select and manage protected areas to support the overall fabric of social and economic development—not as islands of antidevelopment, but rather as critical elements of regionally envisioned harmonious landscapes" (McNeely, 1989:156–57).

This position was influential in the evolution of ICDPs, especially with initiatives such as biosphere reserves. Proponents of the sustainable-use concept have pointed to numerous situations in which people have historically used wildland and wildlife resources without depleting them—citing these cases as evidence that wildlands can be effectively managed in order to harvest their resources. There is little argument against promoting sustained-use activities in areas adjacent to protected areas. But those following the "use" position are increasingly making the case that all areas should be open to some kind of use. Some are advocating that we give up on strictly protected areas altogether (Pimbert, 1993; Janzen, 1994; Wood, 1995).

The entire concept of sustainable use has been based on the ecological fallacy and on utopian thinking, not on science or on a thorough analysis of historical social, economic, and political relationships. Moreover, the concept focuses on use values (presuming that economic use justifies maintaining something) rather than on existence values (the idea that there are reasons other than economic—such as spiritual—why we might value something).

Most of the examples of sustainable use have involved highly regulated and managed ecosystems. Although these uses may be essential as part of an overall mosaic of land-use patterns, there is little scientific evidence that they will be particularly effective in biodiversity conservation. There are many more examples of attempts at sustainable use that have not worked than there are examples that have worked (Brandon, 1994a, 1994b; Freese, 1994). The unpredictability of markets, the level of consumption (local markets versus international ones), and the type and locus of management power are likely to be among the key predictors of success. There has been little understanding of or research on the markets and sustainability of specific products. The types of research that are required to really know if a product is sustainable link an array of disciplines. These include basic population biology (how many seeds or Brazil nuts can be removed before the gene stock weakens?); sociology (what are the social institutions that regulate use and will they remain intact?); marketing (can we transport this easily and efficiently given existing market constraints?); and economics (what is the demand for a given product?). And there has been almost no attempt to examine systematically the notion of use for a given ecosystem. Without realistic and scientifically accurate information, as well as strong sociological and market research, we will only continue to make uninformed guesses about the future of biodiversity under a "use" scenario. (An excellent analysis of sustainable use can be found in Robinson, 1993.)

## Incentives

The assumption is that appropriate sets of incentives can be readily defined and will influence people to conserve biodiversity.

The notion of using economic incentives to achieve biodiversity conservation is not new. But there has been increasing emphasis on the role of incentives in promoting conservation. Perhaps the most influential statement of the role of economic incentives was the IUCN publication *Economics and Biological Diversity* (McNeely, 1988). This report argues: "Economic inducements are likely to prove the most effective measures for converting over-exploitation to sustainable use of biological resources. . . . [C]onservation needs to be promoted through the means of economic incentives" (pp. vii, ix).

McNeely then summarizes the types of incentives and disincentives that have been used to conserve biological diversity; they include direct payments of cash, fees, rewards, compensation, grants, subsidies, credit, and employment. Incentives are important and should not be overlooked in conservation, but there is a big difference between providing incentives to promote the wise use of soil on which a farmer depends for survival, and devising economic incentives or disincentives for keeping people out of parks. McNeely identified a range of projects that had incorporated innovative incentive systems to discourage destruction of wildland resources. But over the long term, these incentives have not necessarily been strong enough to keep people from destructive uses. (Compare, e.g., the discussion of Khao Yai Park by McNeely, 1988, and by Wells and Brandon, 1992; incentives cited by McNeely as working had not, according to Wells and Brandon, led to conservation.)

There are numerous problems in developing appropriate incentives (see chapter 9). Perhaps the most problematic is developing a strong and direct linkage between the incentives and the conservation objective. For example, it is commonly assumed that poor households will switch from illegal, unsustainable, and difficult activities such as poaching to legal activities that generate equal revenue. This assumes that poor households have a fixed income need, and if the households can meet this need through some set of targeted interventions, they will stop their destructive practices. But most poor households want greater income levels—to do better than just holding their own economically—and many households have excess labor capacity during some parts of the year. It is therefore rational for households to continue their illegal activities in the absence of strong deterrence or the presence of a risk they perceive as too great.

It is also difficult to design incentives that work for everybody: should the target be people who are managing resources wisely, or people who are not using them wisely. In many cases, projects may have to devise different incentives for each group, as well as incentives to address each different type of threat. In addition, there is need to figure out what disincentives—usually enforcement activities—are fair, appropriate in the social context, and likely to get people to change their behavior. Field experience shows that identifying both the incentives and disincentives can be an overwhelming task. Compounding the difficulty of developing incentive systems is the fact that many threats are externally generated but acted out locally. For example, even when incentives can be designed for, and with, local people, corrupt government officials or elites can often counter the incentives, especially in the short term and in the absence of strong political will.

Clearly, there is a need to define some economic incentives, to be used appropriately. But it must be understood that using incentives is likely to be extremely difficult and site specific. Furthermore, incentives are usually focused on local people and livelihood concerns. Given the current pattern of forest destruction in which the greatest damage is done by large-scale users, it may be most important to focus economic incentives and disincentives on these groups (Church and Brandon, 1995). For example, it might be most effective to tax cattle ranchers for inefficient land uses or to provide them with incentives to carry out reforestation. Such changes may be difficult politically. And even if they are acceptable, the economic benefits will likely favor wealthy elites—which points again to the difficulty of targeting economic incentives.

## Management

The assumption is that management should be devolved to local control whenever possible.

The successes of some communities, often indigenous ones, in managing resources have gained widespread notice. This has led many to assume that management authority should be devolved to local control whenever possible. Without careful analysis, IUCN adopted the position that it was appropriate to promote the "decentralization of power" to local communities (McNeely, 1989). Unfortunately, this assumption was based on looking at groups who had long-standing mechanisms to manage and control resources.

Maintaining biodiversity in indigenous and extractive reserves depends on the existence of regulatory systems, well-known and locally based technologies, and locally developed management systems. Successful reserve areas have provided benefits principally because they are based on low-density populations that are either intensively extracting from a small area and allowing that area to regenerate or that are extensively using resources collected over a wide area. Successful management systems include incentives for wise management and community cohesion, mechanisms for negotiation and dispute resolution, and ultimately, sanctions of those who fail to use resources well (Brandon, 1994b).

There are three big problems with adopting these approaches widely. First, the systems are appropriate only within their own cultural and ecological context. And even then, these systems may be prone to breakdown, thereby failing to maintain the resource base. Several conditions may make this likely: if there is a substantial increase in the local population; if the area available for exploitation is substantially reduced; if a few commodities increase in value and become more heavily exploited; if social cohesion breaks down with the forces of modernization and market access; or if there are changes in technologies, such as the introduction of guns or chain saws (Redford, 1992; Brandon, 1994b).

Population growth that results from natural increases or migration can cause previously extensive or shifting patterns of resource exploitation to become intensive. Both population growth and resource scarcity (even from natural forces) can lead to changes in community structure, which can undermine local management practices. Loss of community control is usually due to the breakdown of local management institutions, often resulting from forces outside of communities; according to Berkes (1985), "loss of community control over the resource base, commercialization, population growth, and technology change, often occur simultaneously" (p. 204). In Kenya, for example, "where the authority of the elders has been reduced, customary management rules are increasingly difficult to enforce" (Barrow, 1992:6). Clearly, it is unrealistic to think that management systems can be imposed on groups who lack strong social cohesion, where existing social cohesion is undergoing rapid change, or where there is no unified vision of the kind of "development" desired.

Second, in many areas where conservation efforts are being promoted, these (e.g., traditional) management systems do not exist. In many places, projects are working with recent migrants; often the migrants speak different languages or have few common cultural bonds with one another. They are ill equipped to suddenly band together and somehow control resources. Often, there are not even any community organizations with which to work (Seymour, 1994).

Finally, delegating management to communities removes states from the responsibility of assuming what should be a legitimate government function. Some countries are only too happy to turn over parks they care little about to NGOs that will pick up the management and/or funding. But there are serious questions about where the legitimate role for NGOs and communities begins and where the role of the state ends. Furthermore, advocacy groups and local NGOs have strong opinions on what should be done. Although they often claim to speak for local communities, their own constituencies and needs are paramount to those of local peoples (see Wells and Brandon, 1992; Church and Brandon, 1995). These issues have rarely been addressed.

Marshall Murphree (1994) gives an excellent analysis of the current situation:

> [C]urrent literature is full of plans to decentralize conservation management, to "involve" local people in planning, to encourage their "participation" in project implementation and to increase the economic benefits to them arising from resources. However well intentioned, such plans generally fail to achieve their aims of sustainable natural resource management and utilization. "Participation" and "involvement" turn out to mean the co-option of local elites and leadership for derived programs; "decentralization" turns out to mean simply the addition of another obstructive administrative layer to the bureaucratic hierarchy which governs natural resource management. There is no doubt that field-based conservation activities will not get very far unless they clearly identify the stakeholders and their interests, project options, and what the project's impacts are likely be on the different stakeholder groups. Consultation with groups is necessary at a minimum; people's participation in various phases is essential. But getting people to participate in project components is different from the wholesale transfer of decision-making authority to local levels. (p. 405)

## Technology

The assumption is that technical and organizational solutions exist to improve resource management and production activities in areas with great biodiversity.

It is widely assumed that there are packages of technologies that can be adopted—off the shelf—for use in fragile lands. The kinds of needed technologies include everything from natural forest-management techniques, wildlife utilization, and ecotourism, to appropriate agricultural practices and biological prospecting. It has become standard in many conservation projects to (1) provide substitutes for specific resources to which access has been denied; (2) introduce technologies that will provide alternative sources of income to replace those that are no longer available because of the existence of parks; and (3) try to improve or reduce the extent of destructive practices that threaten wildland areas.

Substitutes can be targeted at specific resource uses. For example, if parks were formerly used as a source of fuelwood, establishing woodlots outside the boundaries might provide an adequate substitute. Another example, from the Annapurna Conservation Area of Nepal, is the requirement that visitors bring kerosene into the area to avoid further depletion of the local wood supply. Yet direct substitutes may not always be available outside the protected area, or may not be consistent with the protected area's objectives. For example, if a park represents the only local source of construction materials, medicinal plants, certain fruits, or rare animal species, substitutes probably cannot be provided to supply individuals formerly dependent on these sources. Another problem is that it is often unclear what the appropriate "package" of technology should be to achieve conservation objectives. For example, Sanchez et al. (1982) have argued that intensifying production on tropical lands is possible using high-input (agrochemical) agriculture. Their argument is that these packages will provide the greatest return and thus break the Swidden (formerly slash-and-burn) cycle. This approach varies dramatically from the many small projects—which forego extensive use of chemicals and rely on organic fertilization and integrated pest management techniques—that are most commonly emphasized for such areas.

Even if the technologies exist and there is agreement on their appropriateness, conservationists have often repeated a common failure of rural development projects by neglecting to consider whether the people want the technologies, whether there is adequate social organization to adopt them, or whether they are equally profitable alternatives. For example, in areas where poor peasants chop trees for fuelwood, technical interventions such as biomass methods, solar cookers and collectors, or use of kerosene are often promoted. This perspective generally assumes that such technologies could make major contributions to biodiversity conservation if they were adopted—and perhaps they could. But a wealth of research on how peasants adopt technology shows that peasants generally base their choices on what makes sense from their social, economic, or aesthetic perspective, rather than on consideration of less-critical factors such as preservation for its own sake (Cernea, 1985; Chambers et al., 1989). For a variety of reasons, conservationists have often failed to address all of those needs simultaneously—and this has limited the adoption of numerous technologies.

Projects, whether in the form of media campaigns or household-based strategies, are seen as the appropriate mechanisms for introducing new technologies and the social organization needed to support the new technologies. There is ample evidence, however, that while numerous new technologies have been introduced within the context of biodiversity-conservation projects, rates of adoption have been relatively low (Chapin, 1989; Kiss, 1990; Wells and Brandon, 1992; Church and Brandon, 1995). This finding is similar to the findings of rural development projects that adapting technologies to local preferences can be difficult. It seems clear, therefore, that we cannot rely on technological "fixes" to stop cycles of encroachment.

## Poverty Mitigation and Development

The assumption is that rural poverty-mitigation and development strategies will lead to conservation and maintain biodiversity.

At both the field and policy levels, the links between poverty and environment remain ill-defined. To push at the forefront of the poverty and environment nexus would require defining the policies that can protect biodiversity most effectively while also helping the poor. What is known is that alleviating poverty will not necessarily lead to improvements in biodiversity conservation.

The most common strategy used in ICDPs is to promote social and economic development in communities adjacent to protected areas, with primary emphasis being given to mitigating poverty and fostering community development activities. In this view, the lack of options forces rural people to exploit resources in ways that are unsustainable. Population growth, migration, and declining soil fertility lead to an expansion of the agricultural frontier into wildlands. The only hope for breaking the destructive patterns of resource use is to reduce rural poverty and improve incomes, nutrition, health care, and education. Without breaking these patterns, people will continue to encroach on wildlands.

There are problems with this approach at two levels. First, local people are often the proximate rather than the root cause of threats to biodiversity. Timber concessions, large-scale development activities, and government policies may have a much greater impact than do local people. Second, the development activities promoted have often been unrelated to conservation benefits. For example, one ICDP tried to halt logging in a forest reserve by introducing a large-scale agricultural project in a nearby area that would provide employment and increase local incomes. But the logging continued, in part because there was no explicit link in the minds of local people between growing more and better rice and making more money by logging. Another project targeted livelihood issues among poor resource-dependent peasants. Yet the project ignored the fact that large-scale regional land-use changes—conversions of land for vacation homes and golf courses—placed more of a threat on the park than did peasant activities (Church and Brandon, 1995). Just because conservationists can see the linkage between improved land-use activities and the protection of biodiversity does not necessarily mean that poor people, dispersed over large areas with minimal services, will.

Many ICDP strategies being promoted do not lead to improved conservation or resource use in either parks or adjacent areas. One study of ICDPs found that, despite the tremendous diversity among ICDPs and the prevalence of site-specific conditions, there are similarities in the factors that affect project performance (Wells and Brandon, 1992). Studies of numerous integrated rural development projects, large-scale attempts promoted in the 1970s to improve rural welfare through various policy and project interventions, have revealed similar results (World Bank, 1988; Wells and Brandon, 1992). Much of the failure of ICDPs to achieve their objectives is due to the tremendous complexity in trying to link the social and welfare objectives with biodiversity objectives in settings where the two are not obviously connected.

Further problems are tied to people's desires and to ethical issues. Conservationists have not learned to take consumption patterns and desires into account adequately. Poor people in rural areas want a better standard of living than they have. People in developed countries do not place upper limits on their desired levels of income, and neither do peasants. In general, as long as the opportunity exists to increase income, people are willing to do so—and with increased income comes increased consumption. Indeed, in most cases, this has been the prevailing model of "development." In many cases, having people live in areas such as buffer zones, where the uses are restricted, is equivalent to condemning them to lives of poverty.

## Social

The assumption is that local people are cooperative and live in harmony with one another and with nature.

Redford (1990) has adequately debunked the myth of the "ecologically noble savage." Another important consideration is that biodiversity is less threatened by indigenous groups than by recent migrants to frontier areas. The migrants often have little connection, materially or spiritually, to the land or resources. When indigenous groups are present, there are often serious conflicts between these groups and recent migrants, leading to overwhelming social complexity in getting development or conservation activities going. When indigenous groups migrate from lands that are traditionally "theirs" to other lands, the stewardship that they had may disintegrate when the spiritual attachment is lost, and their actions may make them similar to other peasant migrants (Brandon, 1981).

Conflict over resources and competing types of uses is rampant. Within indigenous communities, there are substantial debates between the young and the old. In many resource-dependent communities, conflicts may arise between men and women, between clans, between neighbors, and so on. Especially thorny may be conflicts between individuals and groups. On a wider scale, competing interests arise over land uses. In the Amazon, for example, rubber tappers, cattle ranchers, and small farmers often come into conflict. In addition, superimposed on the "local" picture are the agendas of outside groups, such as governments and national or international NGOs.

Only recently has the need to identify all the different stakeholders in conservation activities been recognized (Brown and Singer, 1992). It is not uncommon to find 20 or 30 different stakeholders—in government, NGOs, and communities—who are all concerned with what happens in a particular area. But it is rare that there is a "level playing field" so that all voices are heard. Decisions taken at higher levels often have little effect at local levels, and vice versa. And there are many cases in which national decisions have led to increased degradation of resources. The goal is to figure out how to get all the necessary groups effectively represented. To date, there are few models of how this can be achieved for biodiversity conservation.

## THE IMPLICATIONS FOR BIODIVERSITY CONSERVATION

These seven assumptions, taken together, have played an extremely important role in shaping the kinds of activities that are being carried out to achieve biodiversity conservation. Each of the assumptions does hold in some places and in some cases—but we have committed a grave error in letting them become gospel. One reason that the assumptions have lasted is that until recently, there was little interdisciplinary interaction between conservationists and social scientists. Also, until recently, the urgency of preserving biodiversity was largely unappreciated. While this latter oversight is being corrected, there has still not been a good integration of social science into conservation.

To be sure, the degree of interaction has risen remarkably: within the past five years, virtually all major conservation NGOs have incorporated social scientists into their programs. But the common result has been to pull environmental NGOs into rural development concerns, thereby diluting their mission to conserve biodiversity. Rural development per se has been adopted as an objective, which is quite different from the more narrow objective of preserving biodiversity, given existing social and political conditions.

Two important conclusions can be made. First, site-specific judgments that incorporate an understanding of the social, ecological, political, economic, technical, and institutional context must be made for each project and policy. Obviously, clustering these efforts in priority areas is essential to maximize progress. Second, what is being called conservation is no longer conservation. The majority of conservation efforts are in fact large and complicated social programs. Conservation groups lack the expertise to plan or implement such projects alone. By the same token, they should not be expected to shoulder the heavy costs associated with these activities.

We can see how these assumptions have affected biodiversity by looking at their implications on the three types of areas described previously: parks, ICDPs, and locally managed reserves.

### Parks

Four trends are critical in what is happening with parks today. First, parks are not being well maintained. There has been diminished attention to conservation basics, such as collection of basic ecological data, infrastructure development, pragmatic management planning, boundary demarcation, guard training, and enforcement. At the same time, there has been a change in the mission and role of parks: they are now supposed to be "useful" and fulfill a vast number of societal needs. With less funding and attention, greater pressure, and greater management demands, it is not surprising that their effectiveness has declined. Support for parks has dwindled such that there is no consensus among conservationists on the degree to which strictly protected parks are important. It seems as though there are fewer advocates for stronger preservation and reduced use—a position that is now tantamount to being "the bad guy."

Second, there has been little emphasis on improving enforcement. Certainly, no one likes to promote fences, fines, bullets, and barbed wire. But site-specific considerations are often lacking on when it is appropriate to carry out enforcement, and how. Moreover, with the prerequisites for enforcement lacking—such as boundary demarcation and education and outreach activities—policing in fact becomes problematic. Creative approaches for park enforcement in tandem with local communities, especially those that build on existing social structures, are needed.

Third, more parks are needed. As revealed in chapter 3, evidence suggests that current systems of parks will not adequately conserve biodiversity worldwide because they do not represent an adequate array of ecosystems. We do not need to create new parks in all countries, or to create parks simply for the sake of having them. But where needed, parks that are designed and managed based on a thorough understanding of ecological criteria and social realities offer perhaps the best way to maximize the protection of biodiversity.

Fourth, money is needed to buy new parkland and to eliminate, when possible, uses that conflict with biodiversity conservation objectives. For example, voluntary resettlement has rarely been promoted because it is too closely linked with past resettlement programs that have forcibly moved people without compensation. Ironically, in many areas of Asia and Latin America, migrants in search of land would welcome the opportunity to be relocated if the incentives were right. While such incentives are likely to be site specific, a minimum compensation package might include good land, tenure security, and access to markets and community services. It also may be necessary to buy out larger stakeholders, such as logging and mining companies. In Panama, for example, an NGO called ANCON (National Association for the Conservation of Nature) bought 96 percent of the outstanding shares of a gold-mining company that was operating in Darien Park, which effectively ended mining activities. Such buy-outs will take considerable money, time, and creativity to do properly. But, when possible, they are clearly preferable to having enclaves of destructive activity within parks.

## Integrated Conservation-Development Projects

ICDPs are experimental, complex, and generally very expensive undertakings. It is also unclear whether ICDPs, although motivated by humanitarian instincts, are the right vehicle to achieve either development or conservation. There is reason to expect that comparable levels of funding directed at implementing targeted policy reforms might have a better payoff for both people and conservation. The difficulty in implementing ICDPs is well illustrated by the experience in buffer zones. Despite the intuitive appeal of the concept and the frequency with which the term appears in management plans for parks and forests, there are virtually no examples where buffer zones have achieved their goals. Debate has centered on how to draw them, where to put them, and what level of benefits people should derive from them. There is a growing consensus, however, that buffer zones should have park protection as their first priority, with benefits accruing to local people taking a secondary role (MacKinnon et al., 1986; Wind and Prins, 1989; Sayer, 1991). But determining how to achieve this is a challenge.

Brandon and Wells (1992) cite several examples of the difficulties involved in managing buffer zones: "Many potential buffer zone areas are already being exploited, so there is little point in conceding use rights to local people as compensation for lost access to a park" (p. 563). Furthermore, they ask, who decides what levels of uses are allowed? Who is responsible for enforcement? Few protected area management agencies have jurisdiction that extends outside park and reserve boundaries. Therefore, these agencies have no authority to establish or regulate buffer zones in the absence of legislative changes. What happens if the buffer zones begin to deteriorate? What if the buffer zone is maintained but pressure on a park continues? Do you tell people they will have to stop using buffer areas? Who sets the criteria for the level of biodiversity to be maintained in the buffer area? If the community sets them, what happens if the community does not meet its own criteria? If the criteria are established by others, what incentives are there to get people to use the resources wisely? What happens as systems change with time?

Although ICDPs are often more glamourous than projects focused primarily on conservation, they face several problems. First, ICDPs represent a major investment with no certainty of a return. Because of their risky nature, there is no certainty that they will work. Based on experiences with rural development, we also know they will not work unless they are funded for 15 or 20 years.

Second, ICDPs are a poor choice for areas with immediate problems. Because they are so complex, and because it takes them years to establish connections between conservation and development, they are an inappropriate choice in areas where there are declines in biodiversity or high levels of conflict. Stabilizing short-term threats to parks through boundary demarcation, public education, and enforcement are essential first steps to maintain the integrity of areas under threat. These measures can be followed by ICDP approaches if necessary. Such measures can provide strong links between conservation and livelihood, through hiring local people as guards or guides, without the necessity of a full ICDP approach.

Third, ICDPs should be funded by large donors (e.g., the World Bank) and should include both policy-level and project-level components. Without having both in place, there is little chance for success.

It may be difficult for even international conservation NGOs to enter into policy dialogues with the governments in some countries, but the World Bank and other multilateral development banks have the economic muscle to do so. NGOs should be focusing their attention on both the donors and the countries in pressing for policy reforms. Such reform means promoting broad-based sectoral or regional changes, as well as the institutional and political reorientation that will be necessary for ICDPs ultimately to be successful.

## Locally Managed Reserves

Most recent evidence suggests that locally managed reserves are not maintaining the levels of biodiversity conservation that are desired (Robinson and Redford, 1991, 1994). Nevertheless, they represent one of the few mechanisms available that allow local groups to manage biodiversity over the long term. Some species may be lost–and there is no question that biodiversity will be sacrificed. But these kinds of reserves represent one of the best options for concretely linking conservation with development objectives. Reviews of locally managed reserves have stressed either the negative (that they do not achieve biodiversity conservation) or the superlative (that they are great and will save everything). Looking beyond these claims, what is clear is the need to develop local solutions to local circumstances. If there are to be compromises between conservation and development, these are the best places to have them.

Several conclusions can be made regarding locally managed reserves. First, a major worldwide emphasis should be given to identifying indigenous groups with government-conferred land rights. Until recently, little attention was given to solidifying the land or management claims of tribal and indigenous peoples (although Native Lands, a U.S. NGO based in Virginia, now specializes in helping indigenous groups map and gain title to their claims). Once indigenous groups are identified, their efforts to establish land claims should be supported, and discussions should begin on how best to reconcile their needs and the needs of conservation. Priority should be given to groups living near parks. By way of example, The Nature Conservancy, a U.S.-based NGO, is now conducting such interactions with numerous indigenous groups throughout Latin America (Redford, 1996).

Second, technical assistance, support, and training should be given to indigenous groups with secure land rights. A review by Robinson and Redford (1994) of community approaches to wildlife conservation highlights the need that these groups have for increased technical capacity in a range of fields. Transferring methodologies for groups to use in measuring and understanding the conservation impacts of their activities is vital. Also, it is necessary to work with these groups to anticipate conflicting demands for their resources from loggers, colonists, conservationists, governments, and ecotourists. This will provide them with frameworks for management and help them determine what development opportunities are most consistent with their needs.

## FROM RHETORIC TO SUCCESS

There is tremendous variation in the factors that induce people to use and manage biodiversity resources wisely. We know what some of these factors are; traditional groups with high cultural and religious attachment to forests, for example, are more likely to value their preservation than are recent migrants. And we know that some mix of incentives and penalties will probably work in most places. However, we do not know the appropriate mix or how to implement the mix; this can be determined only on a site-specific basis. Moreover, conservationists are far removed from having the political muscle required to achieve the political reorientation necessary for success.

It is tempting to develop a new set of assumptions about the way the world works—that sustainable use is not possible, that management is too complicated, that technologies are lacking, and that conflict is inevitable. But we have to stop the pendulum from swinging to another set of misleading generalizations. Instead, we must pay better attention to the evolving array of projects and policies to conserve biodiversity, and we must quickly translate lessons gained from them into more than optimistic rhetoric. Ultimately, we must use those lessons to guide us on a case-by-case basis in protecting the fragile biodiversity of tropical rain forests.

## NOTES

I am particularly indebted to Kathy MacKinnon and Carter Brandon for their detailed review and comments.

## REFERENCES

Barrow, E. 1992. *Tree rights in Kenya: the case of the Turkana*. Nairobi, Kenya: African Center for Technology Studies.

Batisse, M. 1986. Developing and focusing the biosphere reserve concept. *Nature and Resources* 22:1–10.

Berkes, F. 1985. Fishermen and the "tragedy of the commons." *Environmental Conservation* 12 (3): 199–206.

Brandon, Carter. 1981. *San Quintin: a place apart*. Documentary film on the Tzeltal Maya. Cambridge, Mass.: MIT.

Brandon, Katrina. 1995. The World Bank and the environment. In *Bretton Woods: looking to the future*, 133–42. Washington, D.C.: Bretton Woods Commission.

Brandon, Katrina. 1994a. Trade and sustainable utilization of crocodiles. Unpublished manuscript for the World Wildlife Fund.

Brandon, Katrina, with Amity Doolittle and Jon Kosek. 1994b. Resource tenure and its links to conservation. Unpublished manuscript.

Brandon, K., and M. Wells. 1992. Planning for people and parks: design dilemmas. *World Development* 20 (4): 557–70.

Brown, M., and A. Singer, eds. 1992. *Buffer zone management in Africa: searching for innovative ways to satisfy human needs and conservation objectives.* Synthesis and discussion of workshop in Queen Elizabeth National Park, Uganda. 5–11 October 1990. Washington, D.C.: PVO/NGO/NRMS Project.

Cernea, M., ed. 1985. *Putting people first: sociological variables in rural development.* New York: Oxford University Press.

Chambers, Robert, Lori Ann Thrupp, and Arnold Pacey, eds. 1989. *Farmer first: farmer innovation in agricultural research.* London: Intermediate Technology Publications.

Chapin, Mac. 1989. Ecodevelopment and wishful thinking. *The Ecologist* 19 (6): 259–61.

Church, Philip, and Katrina Brandon. 1995. *Strategic approaches to stemming the loss of biological diversity.* Washington, D.C.: Center for Development Information and Evaluation, U.S. Agency for International Development.

*The Economist.* 1989. How Brazil subsidizes the destruction of the Amazon. 18 March: 69. Reporting on Hans Binswanger, *Fiscal and legal incentives with environmental effects on the Brazilian Amazon* (World Bank Discussion Paper 69, 1989).

Freese, Curtis. 1994. The commercial consumptive use of wild species: implication for biodiversity conservation. (11 Aug. draft). Washington, D.C.: WorldWide Fund for Nature.

IUCN, UNEP, and WWF [World Conservation Union, United Nations Environment Programme, and World Wildlife Fund]. 1980. *World conservation strategy: living resource conservation for sustainable development.* Gland, Switzerland: IUCN, UNEP, and WWF.

IUCN, UNEP, and WWF [World Conservation Union, United Nations Environment Programme, and World Wildlife Fund]. 1991. *Caring for the earth: a strategy for sustainable living.* Gland, Switzerland: IUCN, UNEP, and WWF.

Janzen, D. H. 1994. Wildland biodiversity management in the tropics: where are we now and where are we going? *Vida Silvestre Neotropical* 3:3–15.

Kiss, A., ed. 1990. *Living with wildlife: wildlife resource management with local participation in Africa.* Washington, D.C.: World Bank.

Ludwig, D., R. Hilborn, and C. Walters. 1993. Uncertainty, resource exploitation, and conservation: lessons from history. *Science* 260:17, 36.

MacKinnon, J., K. MacKinnon, G. Child, and J. Thorsell. 1986. *Managing protected areas in the tropics.* Gland, Switzerland: IUCN (World Conservation Union).

McNeely, J. A. 1988. *Economics and biological diversity: developing and using economic incentives to conserve biological resources.* Gland, Switzerland: IUCN (World Conservation Union).

McNeely, J. 1989. Protected areas and human ecology: how national parks can contribute to sustaining societies to the twenty-first century. In *Conservation for the twenty-first century,* D. Western and M. Pearl (eds.), 150–57. New York: Oxford University Press.

Murphree, M. 1994. The role of institutions in community-based conservation. In *Natural connections: perspectives in community-based conservation,* David Western and R. Michael Wright (eds.), Chap. 18. Washington, D.C.: Island Press.

Pimbert, Michel P. 1993. *Protected areas, species of special concern, and WWF.* Gland, Switzerland: World Wildlife Fund.

Redford, K. H. 1990. The ecologically noble savage. *Orion* 9:24–29.

Redford, K. H. 1992. The empty forest. *Bioscience* 42:412–22.

Redford, Kent, ed. 1996. *Traditional peoples and biodiversity conservation in large tropical landscapes.* Rosslyn, Va.: The Nature Conservancy.

Reid, W. V., and K. R. Miller. 1989. *Keeping options alive: the scientific basis for conserving biodiversity.* Washington, D.C.: World Resources Institute.

Robinson, J. 1993. The limits to caring: sustainable living and the loss of biodiversity. *Conservation Biology* 7 (1): 20–28.

Robinson, J., and K. Redford, eds. 1991. *Neotropical wildlife use and conservation.* Chicago: University of Chicago Press.

Robinson, John, and Kent H. Redford. 1994. Community-based approaches to wildlife conservation in neotropical forests. In *Natural connections: perspectives in community-based conservation,* David Western and R. Michael Wright (eds.), Chap. 13. Washington, D.C.: Island Press.

Sanchez, P. A., D. E. Bandy, J. H. Villachica, and J. J. Nicholaides. 1982. Amazon Basin soils: management for continuous crop production. *Science* 216:821–27.

Sayer, J. 1991. *Rainforest buffer zones: guidelines for protected area managers.* Gland, Switzerland: IUCN (World Conservation Union).

Seymour, F. 1994. Are successful community-based conservation projects designed or discovered? In *Natural connections: perspectives in community-based conservation,* David Western and R. Michael Wright (eds.), Chap. 21. Washington, D.C.: Island Press.

Southgate, D., and H. L. Clark. 1993. Can conservation projects save biodiversity in South America? *Ambio* 22:163–66.

Umaña, A., and K. Brandon. 1992. Inventing institutions for conservation: lessons from Costa Rica. In *Poverty, natural resources, and public policy in Central America,* S. Annis (ed.), Chap. 3. Washington, D.C.: Overseas Development Council.

WCED [World Commission on Environment and Development]. 1987. *Our common future.* New York: Oxford University Press.

Wells, M., and K. Brandon. 1992. *People and parks: linking protected area management with local communities.* Washington, D.C.: World Bank, World Wildlife Fund, and U.S. Agency for International Development.

West, P. C., and S. R. Brechin, eds. 1991. *Resident peoples and national parks: social dilemmas and strategies in international conservation.* Tucson: University of Arizona Press.

Western, D., and M. Pearl, eds. 1989. *Conservation for the twenty-first century.* New York: Oxford University Press.

Western, D., and R. Michael Wright, eds. 1994. *Natural connections: perspectives in community-based conservation.* Washington, D.C.: Island Press.

Wind, J., and H. H. T. Prins. 1989. *Buffer zone and research management for Indonesian national parks: inception report.* Bogor, Indonesia: World Bank National Park Development Project, DHV/RIN Consultancies.

Wood, David. 1995. Conserved to death: are tropical forests being over-protected from people? *Land Use Policy* 2 (12): 115–35.

World Bank. 1988. *Rural development: World Bank experience, 1965–86.* Washington, D.C.: Operations Evaluation Department, World Bank.

World Bank. 1992. *Development and the environment, 1992.* Washington, D.C.: World Bank.

# 6

# Biodiversity Politics and the Contest for Ownership of the World's Biota

*Steven E. Sanderson and Kent H. Redford*

In the course of the past decade, biodiversity has become one of the most important concepts guiding conservation and development at the global level. From the 1972 U.N. Conference on the Human Environment, held in Stockholm, to the 1992 U.N. Conference on Environment and Development, held in Rio de Janeiro, concern for biodiversity loss has spawned international treaties, national laws, and community conservation strategies. This concern for biodiversity, however, has not been clearly translated into increased conservation of biodiversity, for a variety of fundamental reasons.

Biodiversity has traditionally been the domain of natural scientists and conservation activists. The first group has focused on the importance of biological diversity for scientific inquiry; the second group has concentrated on the impact of lost biological diversity on social and ecological systems, and has advocated policies to conserve the earth's biota. Increasingly, both groups—and many other constituencies, from sport hunters and fishers to pharmaceutical companies—have fought out the battle over biodiversity in public arenas. The weapons have included national parks and protected areas, species and genetic conservation programs in the field and in other locations such as zoos, private nongovernmental organizations chartered for "genetic prospecting" activities, and integrated small-scale development programs that have a putative conservation side-benefit.

Even as this battle continues, some agreement—if not a consensus—has begun to emerge about biodiversity, which has provided a foundation for common cause among the various constituencies described above. Conservation has become use. The value of biodiversity has come to be determined according to economic criteria alone. Conservation and sustainable development, it is declared, not only *can* go together but are part of the same cloth. Ecological values and economic values are purported to be congruent.

This position masks two disturbing realities that underpin the specific tasks of this chapter. The first reality is that the concerns that fostered the original concept of biodiversity have been surrendered—even forgotten—in the struggle for common ground, to the detriment of science and conservation. The second is that biodiversity and sustainability are far from scientific concepts. Rather, they are political concepts that, because of their scope and complexity, must be treated in light of the most important macropolitical concepts of our time: markets, states, property regimes, and international institutions in a rapidly globalizing political environment.

This chapter therefore contends that biodiversity, a concept that originated in biology, has been wrested from its originators, first by conservation activists and then by economic actors. Conservation activists are committed to a conservation agenda centered on "sustainable use." Economic actors, also advocating sustainable use, are interested in forwarding a conservation agenda that generally does not interrupt the course or abridge the prerogatives of economic growth.

This appropriation of the biodiversity discourse by politically interested actors is significant. It means that the struggle over biodiversity has been converted into a discrete, if complicated, contest for ownership. In the course of making biodiversity "appropriable" for the purposes of sustainable use, society has shifted the ground rules governing biodiversity conservation. Biodiversity has become a political term, and its use now revolves around economic appropriation, not natural preservation. So, as a vivid example, the political discussion of the global biodiversity convention of 1992 was concerned more with ownership than with conservation priorities. The future of biodiversity, accordingly, will depend on the ability of those most concerned with conservation to sharply define priorities for linking life sciences to property regimes. This chapter offers an initial template in that effort.

## CHANGING DEFINITIONS OF BIODIVERSITY

In the past few years, biodiversity conservation has been adopted by countless governments and multilateral organizations as one of their principal planks. In pursuit of this agenda, tens of millions of dollars have been spent, thousands of documents have been crafted, and numerous international treaties have been signed (Redford and Sanderson, 1992). In order to understand the astounding growth in the industry of biodiversity conservation and the ways in which some agreement has been achieved, it is essential to understand the extent to which this accord has been based on the definition of "biodiversity"—or perhaps more accurately, the lack of definition. In fact, the term has been stated as being almost undefinable: "Biological diversity—the ecosystems, species, and genes that together constitute the living world—is *complex beyond our understanding* and valuable beyond our ability to measure" (Ryan, 1992:5, italics added).

The definition of biological diversity has been dynamic, changing dramatically during the past 15 years or so. A brief, incomplete history of the evolution of the definition illustrates this point. Norse (1990) summarizes part of the early history, locating the roots of "biodiversity" in the late 1950s in the work of Hutchinson and MacArthur. In the 1970s, the richness of species was called "natural diversity" by The Nature Conservancy, a U.S. environmental organization, while others used terms such as "genetic resources" or "genetic diversity." In 1980 Thomas Lovejoy, writing in the report *Global 2000* (Barney, 1980), used the term "biological diversity" without definition. The U.S. Council on Environmental Quality's 1980 annual report (CEQ, 1980) used a definition of biological diversity that included the concepts of genetic diversity and species richness.

Despite a lack of definition, the concept proved attractive enough to be the focus of the U.S. State Department/Agency for International Development "Strategy Conference on Biological Diversity," and in 1983 it became the explicit goal of legislation when the U.S. Congress passed the International Environmental Protection Act, which required federal agencies to help conserve biological diversity in developing countries.

In the mid-1980s, the first full definitions of biological diversity were published by Burley (1984), by Norse et al. (1986), and in a publication by the U.S. Congress's Office of Technology Assessment (OTA) entitled *Technologies to Maintain Biological Diversity* (1987). Burley defined biological diversity as including "genetic diversity and ecological diversity, thus encompassing the full range of species and ecosystem dynamics (interactions between species, among species, and between species and their habitats)" (p. 2). This general sense of the term was formalized in the OTA publication, which provided the definition that has remained the one most frequently used or adapted.

In 1988 E. O. Wilson edited the book *Biodiversity*, which, together with an accompanying videotape and teleconference, focused world attention on the concept. At this point, an interesting shift developed in the ways in which the term biodiversity was used. Whereas early definitions clearly included genetic, species, and community or ecosystem components, Wilson and others began to use the term biodiversity as almost synonymous with species richness (Wilson, 1988; Ehrlich and Wilson, 1991). This interpretation was particularly common among taxonomists, who used as ammunition what they knew best—species.

In the early 1990s, as land managers and ecologists became increasingly interested in biological diversity, this trend was criticized: "[I]t is also important to view biological diversity as something more than local 'species richness'" (Probst and Crow, 1991:13). In fact, some ecologists have strongly suggested that the swing toward taxonomic definitions of biodiversity must be countered by giving much more weight to consideration of functional diversity and ecological redundancy (Walker, 1992). Disagreement has reached the point that in an editorial in the journal *Conservation Biology*, Brussard et al. (1992) went so far as to suggest that the term should be replaced with the term "wildlife" because "most people have no idea of what biodiversity means and are unlikely to learn in the near future" (p. 157).

While the community of biological scientists had been arguing about the real definition of biodiversity, many other groups had integrated the term into their objectives. As stated above, biodiversity conservation became required by U.S. federal legislation, and in 1988 and 1991 the U.S. Agency for International Development (USAID) reported to Congress on its efforts to conserve biodiversity (USAID, 1989, 1991). The publicity that the biologists had garnered for biodiversity and the money that national governments, foundations, and multilateral organizations began to spend attracted the interest of these other groups, who demanded standing in the discussion of biodiversity and its conservation. Agricultural scientists, concerned about crop germplasm erosion, became advocates of biodiversity conservation. Ethnobotanists, studying how traditional peoples maintained crop cultivars, joined them. Anthropologists, interested in the loss of traditional cultures, pointed out that if traditional ways were not maintained, biodiversity would suffer. Advocates of endangered species, particularly species in zoos, took up the cry for biodiversity conservation. The pharmaceutical industry, prospecting for new drugs in wild plants, became biodiversity conservationists. And indigenous and traditional peoples, sensing the power of their position, became self-styled defenders of biodiversity.

When the possibility of an international biodiversity conservation treaty was raised, all of these groups demanded to be included. The resulting treaty drafts of what would become the Biodiversity Convention and accompanying documents (e.g., McNeely et al., 1990) reflected the desire to incorporate the agendas of these groups in the definition of biodiversity. This can be viewed in the incorporation of the following statement in the 1991 draft of the treaty convention: "Human cultural diversity could also be considered part of biodiversity. . . . Cultural diversity is manifested by diversity in languages, religious beliefs, land management practices, art, music, social structure, crop selection, diet, and any number of other attributes of human society" (WRI et al., 1992:2).

Defined in these broad terms, everyone is strongly in favor of the conservation of biological diversity; the treaties are signed, the loans approved, and the delegates return satisfied. The success of the biodiversity agenda has been possible only because of the vagueness of the definition. As more parties demanded standing in the discussion of biodiversity conservation, the definition of biodiversity itself became increasingly general.

The broad agreement to conserve biodiversity, defined in such general ways, has had mixed benefits. On the one hand, it has focused international attention on pressing problems and brought together many interested parties. On the other hand, in the process of achieving accord, many very complicated problems have been ignored or glossed over. Biodiversity conservation, as defined internationally, is frequently highly problematic when implemented locally. It is at this local level that conflicts arise between interest groups that could agree only on very broad definitions. Examples of these conflicts include the following:

- At what scale is biodiversity to be measured, and in what units?
- Who has property rights to biodiversity at the local, regional, and international levels?
- To what extent is diversity generated by human selection processes part of "natural" biodiversity?
- Do indigenous people or local people conserve or threaten biodiversity?

Many of these conflicts are an inevitable result of the lack of precise definition of biodiversity and the consequent interpretation of the term in different ways by groups with different agendas. In order to provide a way for such groups to reconcile their disagreements, precise definitions are mandatory. In this spirit, we offer the following definitions:

*Biodiversity* refers to the variety and variability among living organisms, the ecological complexes in which they naturally occur, and the ways in which they interact with each other and with the geosphere. Biological diversity can be measured in terms of different components (genes, species, higher taxonomic levels, communities and biotic processes, and ecosystems and ecosystem processes) and at different scales (temporal and spatial). The different components of biological diversity can be measured in number or relative frequency, or both.

*Genetic diversity* refers to the variability within a species, as measured by the variation in genes within a particular species, population, variety, subspecies, or breed. Much emphasis has been placed on the genetic component of biodiversity for several reasons: (1) protection, particularly of domesticated plants, from genetic vulnerability—that is, from a uniformity that leaves them vulnerable to new environmental or biotic challenges; (2) protection of endangered species; (3) preservation of genes for future use (Ledig, 1988); and (4) allowing all species the ability to respond to future environmental regimes. Referring to the potential value of genetic diversity, Wilson has frequently used the metaphor of the "genetic library" that remains largely untapped by humans (Ehrlich and Wilson, 1991).

Humans have affected the genetic diversity of plants and animals directly through the process of domestication, as well as indirectly through hunting, gathering, habitat alteration, and harvesting (Ryman et al., 1981; Ledig, 1992). Very few people have investigated the ubiquitous effects of human activity on this component of biodiversity.

*Species diversity* refers to the variety of living species at the local, regional, or global scale. It can be measured in a number of different ways that differentially weigh presence and absence versus frequency of different species at a given locality. The species is the unit most commonly used by biologists to categorize the variation of life. It is also the unit best understood by laypeople. As a result of this, coupled with the pioneering efforts of taxonomists, much of the attention on biodiversity has been focused at the species level.

*Diversity of higher taxonomic levels* refers to the variety of organisms within a given region at a taxonomic level higher than the species level, for example, genera, families, orders, and so on. It is clear that the diversity patterns manifested at the species level are by no means always the same as those demonstrated at the generic level and higher. When the objective is to preserve the greatest genetic variation, species from different higher taxa should be selected—that is, the community that contains the most species may not contain the greatest amount of unique genetic information (Platnick, 1991; Mares, 1992). This category is particularly important when comparing marine systems with terrestrial systems, as patterns of variation appear to be different (Thorne-Miller and Catena, 1991).

*Diversity of communities and biotic processes* refers to a group of organisms belonging to a number of different species that occur in the same habitat or area and interact through trophic and spatial relationships. Biotic processes include such processes as pollination, predation, and mutualism.

*Diversity of ecosystems* refers to a community of organisms and their physical environment interacting as an ecological unit. The ecosystem component can be divided into ecosystem types and ecosystem processes. Ecosystem types are bounded communities interacting with the abiotic environment, while ecosystem processes (which comprise one of the critical differences between this component and the previous one) include such things as nutrient cycling and fire. Biodiversity conservation at the ecosystem level seeks to preserve the basic trophic structure of an ecosystem and the patterns of energy flow and nutrient cycling (McNaughton, 1989). It is largely a conservation of properties and processes, not of species or assemblages of species (Walker, 1989).

## THE POLITICS OF BIODIVERSITY

The key issues in biodiversity conservation have come to involve struggle, public policy, "the social question," property rights, the distribution of wealth, and other concepts that are common to the discourse of politics but not to the discourse of ecology or conservation science. The central point of conceiving biodiversity in political terms is obvious but rarely stated: biodiversity conservation is a notion that hovers at the intersection of ecological dynamics and human society—defined by the interaction of economic, cultural, political, and ecological systems.

The course of public policy concerned with biodiversity depends on the outcome of explicitly political conflicts. These include:

- global resource regimes in the international system [e.g., the Biodiversity Convention, the Convention on International Trade in Endangered Species of Flora and Fauna (CITES), and even the "global change" regimes such as the Montreal Protocol and the Framework Convention on Climate Change];
- world agricultural commodity prices, and their impact on the conversion of natural land cover;
- bilateral and multilateral political and economic agreements (e.g., the North American Free Trade Agreement); and
- the greatest political drama—war and peace (e.g., in Burma and Rwanda).

The problem with this political coloration of biodiversity conservation is that very few people trained in the study of politics work on biodiversity conservation and very few people trained in politics understand the science that should underpin conservation efforts. This shortcoming has been affected by two adaptations: the politicization of conservation activists, and the gradual—and accelerating—coming together of conservation and use.

Despite the various hurdles, there are indications that a preservation agenda is emerging, after all (Robinson, 1993). This agenda posits that bigger, more system-driven conservation efforts might offer more benefits than species-driven or small-scale patch conservation (Harris and Silva-Lopez, 1992; Harris, 1993); that not all small-scale economic systems and social systems are equally friendly to the purposes of conservation; that development goes on unabated; and that developing countries are linked to a development design that drives them to manage natural resources intensively, in a manner that will ultimately lead to exhaustion (Sanderson, 1995).

It is also increasingly apparent that in order to succeed at conservation, unusual short-term coalitions are necessary—coalitions between users and preservers, state and society, rich and poor. These coalitions are extremely difficult to put together without sacrificing the principles of conservation activists, and without some parties being bamboozled by politically more experienced actors.

## PROPERTY ISSUES SURROUNDING BIODIVERSITY

In a sense, the real question in the overall debate is, whose biodiversity is it, anyway? Whose genetic material, whose wildlife, whose habitats, whose long-term ecological processes, whose ecological goods and services—and, ultimately, whose biosphere?

These questions may appear facile, but biodiversity is not in reality simply a scientific concept, and biodiversity conservation is not within the purview of science alone. Biodiversity conservation is about governing the world's biota, in the service of openly contested objectives. In turn, the governance of biodiversity depends on its ownership, or at least on claims to stewardship. Biodiversity politics is nothing less than the public struggle over the distribution of the world's biota.

In global politics, there are roughly three kinds of struggles over property in relation to biodiversity conservation: landed property rights, wildlife property (or "fugitive resource") rights, and intellectual property rights. These domains all have different implications for biodiversity conservation, which are worthy of summary treatment here.

There are a number of plausible definitions of property and property rights. For our purposes, property may be described as patterned ways in which people relate to each other in reference to some object (McCay, 1992), and property rights may be described as the claims by one actor of exclusive rights against all other potential claimants in relation to some object (Cohen, 1978; Epstein, 1985; Bromley, 1989). Property is a "bundle of characteristics" (Feder and Feeny, 1991) that includes exclusivity, inheritability, transferability, and enforcement.

Landed property rights are generally assumed to be the most transparent and rigorously applied, though reality falls short of the mark. They may include hunting, passage, gathering, grazing, cultivation, mining, forestry, and other kinds of harvesting prerogatives. The rights to land use may include private, communal, and state governing arrangements, all on the same land (Feder and Feeny, 1991). Landed property rights are certainly the common denominator of property rights, at least as far as conservation is concerned, as they are the most general and comprehensive. It is also true, however, that property rights are altogether inadequate to conservation, and that landed rights, constituting only a part of property rights, are doubly inadequate as a conservation tool.

Wildlife property rights are less commonly treated (Naughton and Sanderson, 1994) but are important, especially when one is considering the relationship between hunting, game species, and the historical development of conservation. Wildlife property rights do not differ from landed property rights as such, except that they are only invested in the land to the extent that it coincides with the animals' spatial behavior. And, quite often, since wildlife regulation began, wildlife property rights have been invested in the state and then ceded in limited degrees to selected individual users or owners. That converts the notion of landed right, viewed through wildlife, into "usufruct" over "habitat," a much more limited concept. Usufruct is the right to use and enjoy the profits and advantages of something belonging to another as long as the property is not damaged or altered in any way.

Competing property claims over wildlife are as old as hunting itself and certainly represent a central dynamic of conservation politics. There is a great deal of variability in the hunting approach to wildlife conservation, as different countries have radically different historical experiences in hunting, different degrees of confidence in allowing their populations to bear arms, and different connections to the international sport-hunting industry, commercial game harvesting, and other determinants of practice and the law governing it. Hunting has left its imprint on the global conservation scene, though, from the International Whaling Convention to CITES, some of which can be reduced to international society's efforts on behalf of the commons to restrict some actors' exclusive rights to harvest and use wildlife or to destroy their habitats (Cohen, 1978).

Intellectual property rights are a new dimension of the biodiversity agenda. In fact, intellectual property is poorly constructed for use in biodiversity, as it mainly revolves around the limited vehicles of patent rights, copyright protection, and trade secrets. Intellectual property law "confers property rights on certain forms of information" (Leaffer, 1990:1). The emphasis in international law has been, until lately, on rights to intangibles in the industrial, scientific, and artistic world, and since World War II and the Green Revolution, on plant genetic improvement. But in the 1990s, a significant point for political dispute, both at the national and international levels, has been the connection of intellectual property rights to conservation. In fact, nowhere has the struggle over control of the world's "biological resources" been more dynamic than in the area of intellectual property and its derivatives.

The arguments about intellectual property rights stem largely from the law governing biotechnology (Greenlee, 1991) and have focused primarily on rights to plant genetic resources. Because of the contentiousness of the issues and the inadequacy of specialized bodies of Western laws to protect a broad spectrum of rights, intellectual property politics now also extends to indigenous rights (King, 1991; Posey, 1991; Simpson and Sejdo, 1992), to higher life forms (Merges, 1988), and to microbial rights (Yoon, 1993; Milstein, 1994). The state, or the public, is mainly brought in where rights are not clearly established or where the provenance of the material in question is alleged to lie in the public domain. The battle is joined by "traditional" claimants to a broad set of rights, proprietary claims by individuals to the products of human innovation, private claims to innovation itself, and state claims to "rights of origin" over the raw material or cultural heritage that generates the intellectual product.

Opposition by the United States to the Biodiversity Convention, which ultimately resulted in President George Bush's refusal to sign, revolved around what his administration and the genetic engineering industry considered to be inadequate provisions for patent protection for U.S. firms engaging in international genetic prospecting. President Bill Clinton's concerns with the same issues induced his administration to introduce property-based stipulations that accompanied its signature of the agreement. And the U.S. Senate, prior to adjourning without a ratification vote in the summer of 1994, showed itself to be ill-disposed toward a strong Biodiversity Convention. In a series of provincial proposals, various bodies in the Senate agreed to recommend the convention only if U.S. national prerogatives in biotechnology and intellectual property were specifically protected, and set the trade-related intellectual property rights standards contained in the General Agreement of Trade and Tariffs (GATT) as the basis for discussing the conservation of biological diversity. The Senate at one point threatened to withdraw if intellectual property rights protections were weak, if a biosafety protocol were agreed to without the advice and consent of the Senate, or if the U.S. contribution to the convention was too expensive. The United States further aggravated developing country partners by declaring the convention's language regarding access to technology to be "voluntary" and not mandatory, as it is understood by other contracting parties, and by insisting on the Global Environmental Facility as the funding mechanism for the convention (Sanderson, 1994).

It is clear that the issue of intellectual property will have a profound effect on the distribution of rewards from investing in biotechnology or discovering "novel" structures and functions for plant and animal material. This will be especially true when the material has been managed previously by an unprotected population (e.g., indigenous peoples) or when the locus of origin of the material is in question.

## THE SCALE OF BIODIVERSITY CONSERVATION

Within the discussion of biodiversity definitions and property regimes lurks an important dimension of scale. As the scope of governance of biotic resources has expanded more and more to the global level in the past two decades, it has collided with conflicting property claims and has generated new claims involving maritime and landed resources, wildlife, and intellectual property. Those conflicting claims are sometimes products of different cultural or economic systems, or of different levels of public intervention, from the village to the nation to the international system itself.

In order to understand the nature and extent of those conflicts, it is necessary to organize the analysis of property rights more thoroughly, especially those involving land and wildlife. Despite the historical and legal depth of landed and wildlife property regimes, the academic literature on conservation regimes tends to treat them as policy instruments, without clarifying the historical, theoretical, and practical political issues arising from different property traditions and claims (Conybeare, 1980; Young, 1989) or virtually without considering the property question in conservation at all (McNeely, 1988; Western and Pearl, 1989). In fact, property regimes determine who has rights over the disposition of wildlife, their habitats, and the ecosystems they inhabit; what strategies for conservation and use societies find optimal; what the distribution of benefits will be; and, importantly, who bears the costs of conservation and use (Bromley, 1982).

A full treatment of these issues is beyond the scope of this chapter. But some summary observations may indicate the complications of scale for deriving property rules for biodiversity conservation.

*First,* biodiversity conservation in current national and international politics is less about conserving the biosphere than it is about who gets to use the biosphere under what property rules, with what allocation of the losses and gains from use.

In June 1992, with great fanfare, the U.N. Conference on Environment and Development (UNCED) issued a biodiversity convention that stipulated in its first paragraphs that the conservation of biodiversity is a global political issue and that nations have sovereign rights over their biological resources (UNCED, 1992). That position responds to a shared sentiment among nations that they should control the present and future rights over the potential economic value of "their" biodiversity. This is an extraordinarily broad claim: it seeks to encompass rights across broad time scales, to allow for future value not presently known, and to make claims that a given quantum of biodiversity is within the purview of a single nation-state.

*Second,* the governance of biodiversity *conservation* is pitched at the national level, while the international organization of *use* creates a superordinate and conflicting set of values.

The current global politics of biodiversity conservation endorses an anachronistic and ineffectual set of property values. The Biodiversity Convention is built around the nation-state, but the argument about rights recognizes the globalization of development and use. With the notable exception of some limited global regimes, biodiversity is governed by nations, without adequate recognition of other important actors. These actors include nonstate national actors, such as indigenous peoples who claim national status; nonnational actors, such as transnational corporations; and international expressions of national actors, such as conservation organizations.

*Finally,* our central point is that the scale at which biodiversity is defined, the biodiversity objectives articulated, and the proprietary regimes chosen to govern it are entrained together, in ways that give significantly different outcomes to different kinds of conservation efforts, according to some identifiable initial choices.

That is, biodiversity is scale dependent. Its meaning varies according to focus on structure, function, or processes, and the management mechanisms brought to bear on conservation are fundamental to the kind of biodiversity conserved.

## THE RELATIONSHIP OF BIODIVERSITY CONSERVATION TO PROPERTY

Referring to the original definitional parameters of biodiversity, the management outcomes sought in conservation efforts, the kind of management mechanism (or institutional venue), and the resource regime itself, we can outline a matrix of biodiversity politics viewed through the property optic. In Figure 6-1, the left side is denominated in terms

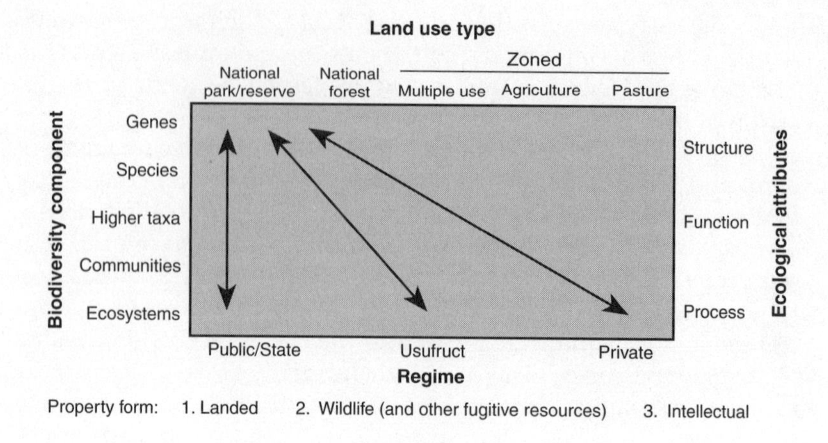


*Figure 6-1* Matrix of biodiversity politics.

of the hierarchical ordering of biodiversity, from genes to ecosystems; the right side in terms of ecological attributes of structure, function, or large-scale ecological processes; the top according to land-management practices; and the bottom according to the property regime and the public or private character of the regime. Additionally, each of the property regimes (public, usufruct, private) can vary according to whether the object is land, wildlife, or intellectual property.

A full discussion of the implications of this diagram is beyond the scope of this chapter. We indicate the possible hypotheses, though.

If we accept a hierarchical approach to biodiversity conservation (see Grumbine, 1992), we might likely opt for a principle of saving the higher level of organization, which in turn will save lower levels. That certainly would suggest conserving biodiversity through an ecosystem approach, rather than a species approach. We could, in turn, modify that by the kinds of ecological attributes identified as priorities.

If we approach biodiversity from the management side, we might suggest that the land-use type and property regimes are scale dependent, such that the larger the land unit, the less likely it is to be in a preserve, and the less likely it is to be held privately. A common-sense hypothesis is that private holdings and usufructs tend to be smaller than publicly managed lands, though this would surely vary according to site. In fact, throughout the world, the historical development of national conservation systems, parks, and other protected areas has created a strange mosaic of land-use types. In some countries, such as Brazil, the older the protected area, the smaller it tends to be, and the closer it is to high-population areas; the newer it is, the larger the size, and the more remote the area from large human populations. In other countries, there are no old parks and protected areas, and the new ones are grafted onto human settlements according to a hodgepodge of conservation strategies, from international to local.

The association of the public sector with protected areas might lead to the conclusion that large public landholdings are a necessary part of any future biodiversity conservation strategy, though the public versus private character of the holding may not be as important as the size of the holding itself.

The relationship among size, conservation, and management flexibility also becomes important to consider in a landscape approach to biodiversity conservation. The nuances of this relationship, too, are beyond the scope of this chapter, but one important consideration is revealed by the observation that protected areas in developing countries typically combine high complexity with small spatial scale. Rarely does a single protected area constitute a significant percentage of a biogeographic province (WCMC, 1992). Only for a limited suite of species are protected areas large enough to cover their home ranges (Harris, 1993). Yet, many protected areas or landscape-scale systems in developing countries are extremely diverse in terms of governance, from the community, to the national government, to the international scene. The beginnings of a multiscale governance strategy for such systems is suggested elsewhere (Sanderson, 1995).

At a more global level, the issue is even more daunting, particularly in light of the struggle over control of the world's biological resources. Evidence from conservation efforts worldwide suggests that there is a definite relationship between the kind of property form (land, wildlife, intellectual) and the kind of biodiversity lost or conserved. More importantly, there are differences in the uses to which biodiversity may be put and the property form and management regimes used to govern that biodiversity.

That is, the social distribution of a country's biota varies according to whether it is collected, identified, preserved, or enhanced by private concessionaires (such as pharmaceutical companies), national nonprofit organizations (such as INBio), national museums, indigenous communities, or other possible contenders. Likewise, the prospect that local people will share in the proceeds from economic exploitation of local biological resources is a direct function of the terms of the contract governing exploitation, and certainly must vary according to the property arrangements.

If we extend the time scale, two additional complications appear. First, intergenerational sustainability becomes a consideration, and the problems of a changing distribution of the costs and benefits of exploiting or conserving biodiversity join the problems plaguing current distribution of the world's biota according to international, national, and local social stratification. "Intergenerational stewardship" presupposes an ability to count on the durability of present-day social arrangements, but little in the political science literature would encourage such confidence, and neo-Malthusian arguments about global environmental degradation (Paehlke, 1989) also discount such optimism.

Second, over relatively short periods of time, issues of origin come into play in assigning rights to biodiversity to individual countries. The idea that a single political unit has an exclusive claim to plant genetic material or any of the other denominators of biodiversity suggests that the biosphere is divided up conveniently according to the political boundaries of the modern nation-state system and that biodiversity is static in place. Conflicting claims over the origins or endemism of exploitable plant and animal material will certainly affect intellectual property claims.

In fact, with little effort, we can topple the very notions on which current rules governing distribution of biodiversity are based. These notions involve the myth of stable ecosystems that permit sustainable exploitation, coupled with the myth of a stable property state as an indicator of rights to biota, and the myth of a stable nation-state system within which the rights to biota might be allocated over the foreseeable future.

To allocate biodiversity for the purposes of conservation and use, we contend that extranational property regimes must play more of a role, in order to implement a preservation agenda that protects large landscapes and integrates surrounding areas of use into a larger preservation agenda (Grumbine, 1992; Robinson, 1993). Second, property mechanisms and biodiversity priorities should emphasize, as much as possible, large-scale ecological processes and ecosystem values (Harris, 1993) rather than single species. And third, within large landscapes, attention must be paid to securing the proprietary rights of small-scale human communities to harvest, but not to the extent of folding traditional use rights into an externally driven property regime (such as international patent rights for indigenous knowledge).

## SOCIAL HISTORY AND NATURAL HISTORY

These principles are not enough, however. Policies for ecosystem management must reflect an appreciation of the natural and social history of the landscape. This is crucial not only for establishing baseline values for ecosystem management but also for determining the legitimacy of human claims to resources in the region. A good example of the importance of natural history and ecosystem knowledge for development policy is found in Africa's Lake Victoria, where a once-diverse system now relies on three species of fish, two of which have been introduced in the past three decades (Kaufman, 1992; Lowe-McConnell, 1993). This lake illustrates the need to observe and respect natural system dynamics—not only fishery abundance in a given span of time, but total ecosystem productivity, ecological function and system organization, and large-scale ecological processes.

As to the social history, the social fabric of biodiversity conservation is not so much a textured cloth as a bunch of holes sewn together. The cross-scale relationships—from local peoples, to national governments, to international systems—are ill-defined and governed by different principles. Even the notions of property we have been discussing are infused with a liberal, Western logic centered on contracts, and the notions often do not resonate with indigenous peoples or longstanding communities in nonmarket economies. There is a great deal to be said for the complaint that biodiversity conservation strategies at the global level are mainly defined by imperial or quasi-imperial notions of legality and rights. It is also clear that the taxonomies chosen to give categories of biodiversity priority are those not of local peoples themselves but of Western science.

If there is a single set of generalizations that might govern social or public choices about property in relation to biodiversity, it is that local institutions should be respected wherever possible; but that local initiatives cannot automatically supplant national or regional system values; and that global biodiversity concerns should be built around landscape-scale system management. To say these things is not the same as devising a strategy to implement them, however. Nor are they necessarily consistent, one with the others. And if the experience of biodiversity conservation to date is an indicator, there is little knowledge about the interaction of ecosystems, local peoples, and large-scale management schemes that is applicable beyond the boundaries of a single site.

## REFERENCES


Barney, Gerald O. 1980. The global 2000 report to the President: entering the twenty-first century. A report prepared by the Council on Environmental Quality and the Department of State. Washington, D.C.: U.S. Government Printing Office.

Bromley, Daniel W. 1982. Land and water problems: an institutional perspective. *American Journal of Agricultural Economics* 64:834–44.

Bromley, Daniel W. 1989. Entitlements, missing markets, and environmental uncertainty. *Journal of Environmental Economics and Management* 17 (2): 181–84.

Brussard, P. F., D. D. Murphy, and R. F. Noss. 1992. Strategies and tactics for conserving biological diversity in the United States. *Conservation Biology* 6 (2): 157–59.

Burley, F. William. 1984. *The conservation of biological diversity: a report on United States government activities in international wildlife resources conservation, with recommendations for expanding U.S. efforts*. Washington, D.C.: World Resources Institute.

CEQ [Council on Environmental Quality]. 1980. Environmental quality 1980: the eleventh annual report of the Council on Environmental Quality. Washington, D.C.: Executive Office of the President. Council on Environmental Quality.

Cohen, Morris. 1978. Property and sovereignty. In *Property: mainstream and critical positions*, C. B. MacPherson (ed.), 153–75. Toronto: University of Toronto Press.

Conybeare, John A. C. 1980. International organization and the theory of property rights. *International Organization* 34 (3): 307–34.

Ehrlich, Paul, and Edward O. Wilson. 1991. Biodiversity studies: science and policy. *Science* 253:758–62.

Epstein, Richard A. 1985. *Takings: private property and the power of eminent domain*. Cambridge, Mass.: Harvard University Press.

Feder, Gershon, and David Feeny. 1991. Land tenure and property rights: theory and implications for development policy. *The World Bank Economic Review* 5:135–53.

Greenlee, Lorance L. 1991. Biotechnology patent law: perspective of the first seventeen years, prospective on the next seventeen years. *Denver University Law Review* 68:127–39.

Grumbine, R. Edward. 1992. *Ghost bears: exploring the biodiversity crisis*. Washington, D.C.: Island Press.

Harris, Larry D. 1993. Some spatial aspects of biodiversity conservation. In *Our living legacy: proceedings of a symposium on biodiversity in British Columbia*, M. A. Fenger, E. H. Miller, J. A. Johnson, and E. J. R. Williams (eds.), 97–108. Victoria, B.C.: Royal British Columbia Museum.

Harris, Lawrence, and Gilberto Silva-Lopez. 1992. Forest fragmentation and the conservation of biological diversity. In *Conservation biology*, Peggy Fiedler and Sunbodh K. Jain (eds.), 197–238. New York: Chapman and Hall.

Kaufman, Les. 1992. Catastrophic change in species-rich freshwater ecosystems: the lessons of Lake Victoria. *BioScience* 42:846–58.

King, Steven R. 1991. The source of our cures. *Cultural Survival Quarterly* (Summer): 19–22.

Leaffer, Marshall A., ed. 1990. *International treaties on intellectual property*. Washington, D.C.: Bureau of National Affairs.

Ledig, F. T. 1988. The conservation of diversity in forest trees. *BioScience* 38 (7): 471–79.

Ledig, F. T. 1992. Human impacts on genetic diversity in forest ecosystems. *Oikos* 63:87–108.

Lowe-McConnell, Rosemary H. 1993. Fish faunas of the African Great Lakes: origins, diversity, and vulnerability. *Conservation Biology* 7:634–43.

McCay, Bonnie. 1992. Everyone's concern, whose responsibility? The problem of the commons. In *Understanding Economic Process*, Sutti Ortiz and Susan Lees (eds.), 199–210. Monographs in Economic Anthropology, No. 10. Lanham, N.Y.: University Press of America.

McNaughton, S. J. 1989. Ecosystems and conservation in the twenty-first century. In *Conservation for the twenty-first century*, David Western and Mary C. Pearl (eds.), 109–20. New York: Oxford University Press.

McNeely, Jeffrey. 1988. *Economics and biological diversity: developing and using economic incentives to conserve biological resources*. Gland, Switzerland: IUCN (World Conservation Union).

McNeely, J. A., K. R. Miller, M. V. Reid, R. A. Mittermeier, and T. Werner. 1990. *Conserving the world's biological diversity*. Gland, Switzerland: IUCN (World Conservation Union).

Mares, Michael A. 1992. Neotropical mammals and the myth of Amazonian biodiversity. *Science* 255:976–79.

Merges, Robert P. 1988. Intellectual property in higher life forms: the patent system and controversial technologies. *Maryland Law Review* 47:1051–75.

Milstein, Michael. 1994. Yellowstone managers eye profits from hot microbes. *Science* 264:655.

Naughton, Lisa, and Steven Sanderson. 1994. Property, politics and wildlife conservation. Manuscript.

Norse, Elliott. 1990. *Threats to biological diversity in the United States.* Washington, D.C.: Office of Policy Planning and Evaluation, U.S. Environmental Protection Agency.

Norse, E. A., K. L. Rosenbaum, D. S. Wilcove, B. A. Wilcox, W. H. Romme, D. W. Johnston, and M. L. Stout. 1986. *Conserving biological diversity in our national forests.* Washington, D.C.: The Wilderness Society.

Office of Technology Assessment, U.S. Congress. 1987. *Technologies to maintain biological diversity.* Washington, D.C.: U.S. Government Printing Office.

Paehlke, Robert C. 1989. *Environmentalism and the future of progressive politics.* New Haven, Conn.: Yale University Press.

Platnick, N. I. 1991. Patterns of biodiversity: tropical vs temperate. *Journal of Natural History* 25:1083–88.

Posey, Darrell. 1991. Effecting international change. *Cultural Survival Quarterly* (Summer): 29–35.

Probst, J. R., and T. R. Crow. 1991. Integrating biological diversity and resource management. *The Journal of Forestry* 89 (2): 12–17.

Redford, Kent H., and Steven E. Sanderson. 1992. The brief, barren marriage of biodiversity and sustainability. *Bulletin of the Ecological Society of America* 73:36–39.

Robinson, John. 1993. The limits of caring: sustainable living and the loss of biodiversity. *Conservation Biology* 7:20–28.

Ryan, John C. 1992. *Life support: conserving biological diversity.* Worldwatch Papers, No. 108. Washington, D.C.: Worldwatch Institute.

Ryman, N., R. Baccus, C. Reuterwall, and M. H. Smith. 1981. Effective population size, generation interval, and potential for loss of genetic variability in game species under different hunting regimes. *Oikos* 36:257–66.

Sanderson, Steven. 1994. North-south polarity in inter-American environmental affairs. *Journal of Inter-American Studies and World Affairs* (Fall): 25–47.

Sanderson, Steven. 1995. Ten theses on the promise and problems of creative ecosystem management in developing countries. In *Barriers and bridges to the renewal of ecosystems and institutions.* L. Gunderson, C. S. Holling, and S. Light (eds.), 375–390. New York: Columbia University Press.

Simpson, R. David, and Roger Sejdo. 1992. Contracts for transferring rights to indigenous genetic resources. *Resources* (Fall): 1–6.

Thorne-Miller, B., and J. Catena. 1991. *The living ocean: understanding and protecting marine biodiversity.* Washington, D.C.: Island Press.

UNCED [United Nations Conference on Environment and Development]. 1992. Biodiversity convention. Typescript.

USAID [U.S. Agency for International Development]. 1989. *Conserving tropical forests and biological diversity: 1988–1989 report to Congress on the USAID program.* Washington, D.C.: USAID.

USAID [U.S. Agency for International Development]. 1991. *Tropical forests and biological diversity: 1990–1991 report to Congress on the USAID program.* Washington, D.C.: USAID.

Walker, B. H. 1989. Diversity and stability in ecosystem conservation. In *Conservation for the twenty-first century*, D. Western and M. Pearl (eds.), 121–30. New York: Oxford University Press.

Walker, B. H. 1992. Biodiversity and ecological redundancy. *Conservation Biology* 6 (1): 18–23.

WCMC [World Conservation Monitoring Centre]. 1992. *Global biodiversity: status of the earth's living resources.* London: Chapman and Hall.

Western, David, and Mary Pearl, eds. 1989. *Conservation for the twenty-first century.* New York: Oxford University Press.

Wilson, Edward O., ed. 1988. *Biodiversity.* Washington, D.C.: National Academy of Sciences Press.

WRI, IUCN, and UNEP [World Resources Institute, World Conservation Union, and United Nations Environment Programme]. 1992. *Global biodiversity strategy: guidelines for action to save, study, and use Earth's biotic wealth sustainably and equitably.* Washington, D.C.: WRI, IUCN, and UNEP.

Yoon, Carol Kaesuk. 1993. Counting creatures great and small. *Science* 260: 620–22.

Young, Oran R. 1989. *International cooperation: building regimes for natural resources and the environment.* Ithaca, N.Y.: Cornell University Press.

# 7

# User Rights and Biodiversity Conservation

*Marie Lynn Miranda and Sharon LaPalme*

The management of tropical forests has evolved considerably during recent decades. In the 1970s, the colonial and postindependence emphasis on maintaining large plantations and maximizing timber production gave way to a dual emphasis on revenue generation and social forestry. More recently, the international community, including developing countries themselves, has begun to recognize the important environmental services provided by tropical forest resources, including water quality, soil retention, biodiversity, and microclimate and macroclimate regulation.

Just as the prevailing view of appropriate objectives for tropical forest management has changed, so has support for the devolution, or transfer, of rights to local people. Under the previous forest-management paradigm, which stressed revenue generation and social forestry, governments and international aid agencies encouraged nationalization of forests and the gazetting of land into systems of state forest preserves. This served, perhaps unintentionally but nevertheless forcefully, to restrict the rights of locals. But as the relationship between the landless poor, indigenous groups, and the forest resource came to be better understood, more consideration was given to allowing communities to retain or gain customary and/or legal rights to the forest resource. Now, however, by adding the protection of environmental services to the management paradigm, the effects on the devolution of rights to local people are much less clear. On the one hand, some would argue that the only way to vest locals in the maintenance of the forest resource is to give them specific, income-enhancing rights to its use. On the other hand, examples abound of local populations who have exploited the forest resource in ways that are not sustainable, destroying fragile ecological relationships and degrading the biodiversity of the area in the process.

The support for devolution of rights has waxed and waned over the years, with its popularity dependent on both international politics and the world economy. The question of whether to devolve rights becomes especially complicated when considering the fate of protected areas in the tropical developing world. Within the protected areas themselves, user rights exercised by local people either can be relatively benign or can have devastating effects on the local ecosystem. Even if a country has declared areas to be strictly protected, it must still decide what to do about devolving rights in the surrounding areas. Clearly, successfully defending protected areas depends on providing local people with sufficient opportunities to earn income and achieve an acceptable quality of life.

After decades of discussion, rhetoric, and experimentation, it has become clear that there is no single best answer to the question of devolution. Certain approaches, however, may be better suited to the local political, economic, cultural, and ecological context. We begin this chapter by providing definitions and historical background on user rights. We identify population pressures and equity considerations as the primary issues shaping the devolution debate. We explore some of the common conflicts surrounding user rights and tropical forest resources. We highlight three case studies of projects that have taken different approaches toward devolution with varying success. Finally, we draw on the earlier analysis to determine when and where it makes sense to devolve user rights, if the objective is to protect biodiversity.

## DEFINING USER RIGHTS

### Types of Rights

The rights to use forest resources in the developing world are covered by a complicated mixture of formal and informal rules. These rights, of course, are not absolute. Different ownership categories (government owned, privately owned, communally owned, open access) have different user rights and entitlements for owners and formal or informal user rights for nonowners. The rights vary from country to country and from user group to user group, but the most relevant rights include (Schlager and Ostrom, 1992)

- access–the right of entry into a defined area;
- withdrawal–the right to remove or use products of the resource;
- management–the right to regulate use and improve or change the resource;
- exclusion–the right to grant access to the resource and to define transfer rules for access; and
- alienation–the right to sell or lease management and exclusion rights.

The rules outlining these user rights may be de jure, defined and enforced by the government with legal recognition, or de facto, defined and enforced by the resource users. Complementary, overlapping, or conflicting de jure and de facto rights sometimes exist within a single resource area, and both governments and resource users vary in their ability to enforce use restrictions. The source of authority for de jure user rights may emanate from deeply held community views of the forest resource, as well as from government policy or legislation.

## Historical Background

User rights to land in the developing world are a collage of European-derived property laws applied to countries and regions during colonization (Adeyoju, 1976) and customary rights or traditional tenure systems characteristic of indigenous societies that survived colonization or were reinstituted after independence (de Saussay, 1987). Group ownership is common in some regions of Asia and Africa (Adeyoju, 1976). In Latin America, communal ownership has been replaced essentially by private or state ownership (Millon, 1955; Poore et al., 1989).

Governments have commonly claimed state ownership of forest land, creating protected forests, production forests, or forest reserves (Gibbs and Bromley, 1989). Frequently, state ownership of forest land replaced customary common property tenure regimes. Colonial governments often assumed that communally owned land was ownerless, or at best poorly managed, and appropriated these lands. As expressed by Troup (1940): "If State ownership of forests is considered necessary in European countries where the importance of forestry is well understood, it is even more necessary in the Colonies, where the inhabitants have as yet developed little or no 'forest sense,' and are often antagonistic to forest preservation in any form" (p. 281).

Troup continues by lamenting the forest destruction in British Honduras and pointing to the success of state ownership of teak forests in Burma. Regarding the setting of policy for Africa, he cautions that any forests set aside for local control and use should be "forests of comparatively small area and no great timber value" (p. 282). Despite their "protected" status, however, state ownership was usually motivated by a desire to control timber resources and revenue rather than to protect "nature."

India embraced this same notion of state forests. The British colonial government recognized three types of forest land: reserved forests (the most regulated under the authority of the Forestry Department), protected forests, and unclassified or village forests in which some utilization by surrounding communities was allowed (Commander, 1986). The government redefined customary rights as privileges to be granted at the discretion of local rulers. Tribal communities reliant on the forests for food, fuel, fodder, tools and building products, spiritual significance, and village social and political stability were often devastated by these policies.

By India's independence in 1947, state-held forests totaled 99,400 square miles (Guha, 1983b), well over 40 percent of the country's total forest area (Commander, 1986). Colonial property-rights policies and their effects on the community structure, functioning, and customary rights remained in effect after independence. India's national forest policy, enacted in 1952, states that the colonial forest policy (specifically the 1894 act) "constitutes the basis for the forest policy of India up to this day" (Guha, 1983a:1888). Despite this long and entrenched history of state-held forest land, India met with only limited success in protecting its protected areas. First, timber production was emphasized without consideration of its impact on environmental services like biodiversity. Second, access rights to the forest were often handed out as political favors, with little consideration given to the needs and desires of the local poor and landless. And third, despite competent personnel, the Forestry Department was unable to diffuse the pressure from a quickly growing population seeking fuelwood, fodder, forest products, and arable land within the state preserves.

Colonialism's effects on native peoples and forest lands have been equally powerful in the Western hemisphere and exhibit the same colonial and postcolonial trends. In Brazil, the exploitation of forest lands and the usurpation of tribal communities' land-use rights began in the 16th century. The treatment of the Indians' land as commercial reserves for timber, minerals, and water to fuel economic development continued for centuries. The rights of Indians to their traditional lands and the recognition of their social organizations, customs, languages, beliefs, and traditions were only formally recognized in June 1988 with the adoption of a new Brazilian constitution (Ecumenical Center, 1989). While this new policy addresses important equity concerns, the implications for biodiversity preservation in the protected areas are unclear (see chapter 4).

The gazetting of land into state forest preserves or the privatization of land through various agricultural, resettlement, or forestry projects has been widely supported directly or indirectly by multilateral lending agencies—especially during the immediate postcolonial period. The Polonoroeste project in Brazil, the transmigration program in Indonesia, and the cattle-ranching project in Botswana are examples of lending projects based on inappropriate land-use patterns and lack of consideration of customary land-use practices (Aufderheide and Rich, 1988). In many ways, these projects combine the worst aspects of each course of action: they promote a type of development that undermines local populations by usurping customary user rights and that degrades biodiversity by encouraging large influxes of people into fragile areas.

This philosophy favoring privatization or nationalization was echoed by Western academics and policy experts and was perhaps best exemplified by Garrett Hardin's seminal 1968 article. In it, Hardin describes the "tragedy of the commons"—a situation where a failure to control individual impulses results in long-term degradation of open-access resources. According to this argument, the solution rested with privatization or nationalization in order to set up what were considered more appropriate land-use incentive schemes. The decision between privatization or nationalization usually depended on the sophistication of market structure and administrative capability in the country, as well as on the ideology of the political leadership.

Much of the dialogue regarding privatization and nationalization took place long before biodiversity had taken its current place on the international agenda. The gazetting of land into state-held forest preserves often led to a breakdown in the informal use rules that had previously mitigated against excessive pressure on the forest resource. Government forest agencies often lacked the information, personnel, and financial resources required to design, implement, and enforce sustainable management plans. In countries that emphasized privatization, private and individual property rights created little incentive to provide for environmental services, like biodiversity, whose benefits spill over beyond the actual owner. In other countries, state-held forests were viewed with antagonism as vestiges of the colonial regime. So neither the nationalization nor the privatization route did much to preserve and protect biodiversity. In fact, both approaches most often emphasized the use values of the forest resource, with little consideration given to the inherent biodiversity value. Thus, these state-held forest preserves or protected areas were often designated in order to protect income opportunities rather than biodiversity.

## ISSUES SHAPING THE DEVOLUTION DEBATE

Devolving or not devolving user rights to local people is bound to have a tremendous impact on the viability of protected areas. Devolution decisions in the tropical developing world are especially complicated because they must be made within the context of growing population pressure and troublesome distributional inequalities. These issues call for creative arrangements to incorporate the concerns of all stakeholders, in an equitable manner, in situations whose dynamics are changing almost too fast to comprehend.

### Population Pressure

In developing countries, both large rural and urban populations depend on forest resources for their daily subsistence and energy needs. As a result, the demands on tropical forest resources can easily exceed the ecosystem's carrying capacity and disrupt formerly sustainable land-management and user-rights schemes. Similarly, population pressure may drive incursions into protected areas, at the expense of preserving biodiversity.

In 1990, global population was roughly 5.3 billion (WRI, 1992). Projections indicate that by 2025, global population will grow to 8.5

billion—an increase of 60 percent during a 30-year period (WRI, 1994). Most of this growth will occur in the developing world, and the relatively large proportion of young people globally means that this high level of growth can be expected even if countries institute aggressive family-planning policies.

Key population indicators for countries with significant remaining areas of tropical forests provide mixed signals; these countries include Brazil, Bolivia, Ecuador, Peru, Colombia, Venezuela, Guyana, Suriname, Zaïre, Congo, Gabon, Cameroon, Central African Republic, Malaysia, Indonesia, Papua New Guinea, Thailand, Myanmar, and the Philippines. The average annual population change (in unweighted averages) stood at 2.42 percent during the 1980–1985 period, and projections for 1990-1995 and 2000–2005 fall to 2.19 percent and 1.86 percent, respectively (WRI, 1994). The total fertility rate, however, remains very high. In 1970–1975, the total fertility for these countries (in unweighted averages) stood at 5.52; by 1990–1995, this figure had only fallen to 4.09 (WRI, 1994).

The impact of rapid population growth on the environment will, of course, vary from country to country. At the very least, population increases will elevate the demand for basic resources such as land, energy, and water. Even with a clear and coordinated population-control policy, growing numbers of people will likely place increasing pressure on tropical forest resources and the biodiversity they house.

Population pressures result from both annual population growth rates in many developing countries that average at least 2 percent (WRI, 1992) and population migrations, either voluntary or compulsory, to rural tropical lands. Colonization programs to relieve population pressures in urban areas and voluntary migration often bring people onto land unable to support them. Buffer-zone or extractive-reserve projects often attract more people than they were originally designed to sustain.

Population pressures, however, are closely entwined with problems of poverty, inequitable access to land and other resources, and general development and economic policies. Poor rural and urban populations usually have few alternatives to using wood for fuel. In fact, nearly 2 billion people depend on fuelwood for cooking and heating (WRI, 1994). Rural communities are often relegated to marginal forest land or possess few user rights. Increased numbers of individuals dependent on the forest resource further degrade marginal lands and encourage forest use and exploitation in other forest areas that may be designated for conservation or protection.

In addition to acknowledging their countries' growing numbers of people, policymakers need to be realistic about those people's changing aspirations. Education traditionally has raised expectations for quality of life. In addition, while policies that raise people above the poverty line are sorely needed, it is unrealistic to expect them to be satisfied with some minimal standard of living. As a result, programs originally designed to *substitute* for income flowing to the rural poor through exploitation of forest resources are often viewed by the rural poor as *additional* to the destructive activities (see chapter 5). In the end, very little of the pressure on biodiversity resources is relieved.

General economic development will have a tremendous effect on the long-term impacts of rapid population growth on tropical forest resources. The more successful countries are in creating employment opportunities elsewhere, the more likely protected areas will remain intact. To the extent that the developing world now accounts for close to 95 percent of population growth, however, job creation of this magnitude represents a formidable task.

As population pressures mount, governments and communities may be tempted, either officially or unofficially, to use the remaining forest resource—including protected areas—as an outlet valve for the landless or unemployed. Given the fragility of much of the remaining tropical forest ecosystem, this can be at best a very short-term solution. Individuals, households, communities, and governments must adjust the full range of their behaviors to account for the growing numbers of people in the tropical developing world.

See Box 7-1 for examples from Niger and China that highlight the destructive relationship between population pressure and forest resources.

## Equity

The second major challenge in forest use and management is the struggle to achieve equitable outcomes. In developing countries, formal and informal user rights and distributional equity are often in conflict. When inequity in land rights occurs, it often follows preexisting divisions between societal groups (e.g., peasants versus rural or urban wealthy; indigenous peoples versus colonists; nonindustrial agriculturalists and forest users versus agroindustries and timber companies; and men versus women). Furthermore, inequity in user rights is frequently associated with resource degradation and has resulted in wide-scale environmental destruction. Individuals disenfranchised by such an inequitable system may have little choice but to rely on protected areas to enhance their income.

Thiesenhusen (1991) reviews numerous accounts of inequitable distribution of land resources leading to environmental degradation in Latin America. The *latifundio-minifundio* (large farm–small farm) arrangement is pervasive throughout the region and has resulted in controversial landholding patterns. For example, in El Salvador up to 92 percent of the farms account for only 27 percent of the land area and are too small to satisfy the minimum needs of a family. In contrast, 2 percent of the farms account for 50 percent of the area and are large enough to employ outside workers. These large farms commonly oc-

***Box 7–1*** Population pressures on the forest resource: examples from Niger and China

The landless poor, often encouraged by explicit government policy, are rapidly converting forested land to agriculture, both within and adjacent to protected areas. Two examples from Niger (see Obi, 1963) and China (see Menzies, 1988) highlight the impact that population pressure can have on forest resources.

### Niger

Niger's recent history can be divided into three phases based on resource characteristics: resource abundance (1884–1935), equilibrium (1935-1974), and relative scarcity (1974 to present). In the first phase, the locus of control was at the village level. Joint use of land was common, although fencing was sometimes used to exclude individuals from separate groups or tribes. Ownership was separate for two food species—baobab and date palm were privately owned, while all other species were open to free and equal access. During this period, villagers did not perceive a need for active forest management.

In the second phase, the colonial government subsumed control over property and created state forests to be regulated by the Forest Service. Private ownership of the baobab and date palm was abolished. These two species and 13 others came under the regulation of the Forest Service. With enforcement of the new policy, villagers responded in ways that would eventually exacerbate the resource scarcity problems. Their actions included destruction of seedlings of protected species, surreptitious cutting of mature individuals, and abandoning management of natural regeneration.

While increasing population pressure may have required the government to expand its role in the forest sector, the means of intervention were problematic. The government failed to consider the historical relationship that villagers had to the two major food tree species and the important income flows from these species. By changing both the economics and politics of traditional resource use, the government created a population that was unwilling to cooperate with the forest-management plan. The Forest Service, in turn, did not have the personnel or financial resources required to enforce use restrictions.

Niger's policies—though intended as a response to growing population pressures—were unable to respond to the changing needs of the growing local population. In the end, none of the resource users had the incentive to explicitly consider long-term environmental impacts—a situation that resulted in adoption of unsustainable forest practices and helped usher in the third phase, resource scarcity. If implemented near or adjacent to protected areas, spillover pressure would likely have led to rapid degradation of the reserved areas.

*Box 7–1* (continued)

China

Customary law in China, in contrast to Niger, resulted in more sustainable practices even under growing population pressures. Tenants on China's wildlands who wanted to plant trees could obtain contracts that accommodated long tree-growing cycles. This was in contrast to the usual form of tenancy on agricultural land under crop production, which typically lasted from three to five years. Tenants could also sell their usufruct rights—that is, their right to use land owned by someone else—while the land still remained the property of the landlord. Tenancy agreements became very complicated if tenants cultivated subsistence crops before the forest canopy closed or grew fungi and mushrooms in clearings and on logs and stumps. Communities and villages also gained usufruct rights on common lands—often with some sort of supervisory or management group that acted to schedule harvesting and maintenance and to ensure sustainability.

The approach in China provided a regular source of income to the landless poor by extending links to valuable product markets. While this approach is utilized in very limited areas, it does lay out a specific set of user rights held by individuals both across the community and across time. The customary law explicitly acknowledges the income that flows from particular user rights and the implied distributional impacts. By acknowledging the important income flows available to resource users who are poor, this arrangement of user rights managed to engage the cooperation of the local community and ease some distributional concerns. The system acknowledged the traditional relationship between locals and the land while simultaneously providing some limits on excessive use. This, in turn, created an incentive for the landless poor—who in other countries would have converted the land from wildland to agricultural land (legally or illegally)—to keep land under forest cover. This approach combines some of the strengths of the private and communal systems and encourages the landless to take a longer view of the forest resource.

cupy the most productive, flattest, and most easily cultivated lands (Leonard, 1987). The imbalanced distribution of land also results in large numbers of landless people. For example, in Brazil the landless are often unable to escape poverty and unemployment in urban areas and migrate to the rural frontier to claim land in rain forests. The land is burned and cleared, cultivated for several seasons, and then sold for pasture.

Income class and status often affect the de facto and de jure user rights held by individuals and community groups. In Costa Rica, for example, landowners must apply for a special permit from the state forestry agency in order to undertake any tree-harvesting activity. This process is often slow and cumbersome for the landowner. In addition, many small landowners claim that permit applications from wealthier and more powerful landowners are processed more quickly and more favorably. Costa Rica's policy of separating land and tree tenure has created uncertainty over the specific user rights held by landowners. In addition, the policy has created a perception of unequal access to the bureaucracy issuing harvesting permits. Many landowners, sensing that they cannot control the income available from forest-based activities unless they are in some privileged minority, do not find it worthwhile to keep their land under forest cover over the long term. This may account, in part, for why Costa Rica has such a high deforestation rate outside the national parks.

Class-based inequities are also apparent in Thailand. Thailand maintained a common property system based on usufruct rights (the right to use land belonging to someone else so long as the land is not damaged or altered) contingent on continuous use; that is, if land lay fallow for more than three years, rights were forfeited. As agriculture became a profitable commercial venture, land values appreciated and titling was introduced. Titling procedures went through many variations but continually suffered from inadequate recordkeeping, duplication, and lack of standardization. Until titling procedures were improved, farmers were not guaranteed secure property rights, which hindered the adoption of more intensive agriculture practices. Differential access to titling procedures was evident as the poor were frequently unable to obtain full documentation. Government officials and elites used the system to evict homesteaders, and non-Thai minorities were often dominated in mountain areas (Bruce and Fortmann, 1988).

Finally, gender often defines user rights for both land and forest resources. In the developing world, women's rights and interests are defined most strongly by the needs of the household; women collect fuelwood, food, fodder, and plants useful for medicine or cooking, and they are often cultivators of agricultural crops. Men's interests are defined more as income earners; they harvest timber and building poles, sell fuelwood, and often do not rely on many secondary and tertiary forest products (Gregersen et al., 1989). When inequities exist in land-use rights or the quantity or quality of forest products allowed to be used, it is frequently to the detriment of women. Once areas originally used for fuelwood and fodder are degraded, converted, or cordoned off, women must spend more time collecting these products and less time cultivating crops. The consequences frequently include decreases in the nutritional health of family members, the rearing of more children to help meet additional labor requirements, and deforestation of more lands in search of adequate fuelwood (Jacobson, 1992).

User rights influence an individual's standard of living by defining who has access to forest products. This is particularly true of women who are most commonly the gatherers of forest products. Tenure determines the distribution of benefits resulting from reforestation or afforestation (planting trees in new areas), extraction of minor forest products, and timber harvests. These benefits often do not accrue to women, those with seasonal tenure rights, or the landless. These disenfranchised groups, lacking the political clout to gain user rights elsewhere, often must resort to making illegal incursions into protected areas.

Both population pressure and distributional concerns must be addressed in an environmentally and socially sensitive and economically sensible manner. In poor communities, access to the forest can mean substantial improvement in quality of life. As a result, conflict has often arisen over whose concerns take precedence in managing the forest resource.

## COMMON CONFLICTS SURROUNDING FOREST USER-RIGHTS ARRANGEMENTS

The historical legacy of property-rights arrangements has indelibly marked the allocation and distribution of current user rights. This, in turn, has tremendous impact on the degree to which groups and individuals respect the use restrictions that usually accompany the establishment of protected areas. Security in user rights can be accomplished through a variety of mechanisms, including privately held property rights, state-held property rights, or communally based property rights. All three arrangements, however, require clearly defined and enforced use rules. The uncertainty associated with insecure tenure creates perverse incentives for individuals to exploit the forest resource at unsustainable rates. In addition, it often causes conflicts among the different potential users of the resource.

### States Versus Traditional Users

Governments of developing countries often push for the creation of state forest preserves, which fosters the exclusion of traditional users. The new systems of management may ignore the ecosystem knowledge and the deeply ingrained customary views of the forest resource implicit in many of the management approaches used by indigenous cultures. Designating land as a protected area often creates uncertainty among traditional users regarding their right to continue longstanding forest-based activities. As a result, individuals are often motivated to exploit the forest resource at unsustainable rates in order to take advantage of the time lag between nationalization of forest lands and development of a clearly enforced national management policy.

One example comes from Nepal. In 1957, the government replaced de facto communal forest rights with de jure state rights in an effort to halt deforestation. However, the government proved incapable of enforcing this new property regime. The activities of the local population were no longer restricted by community forest-management policies, and the locals also realized that the government's management and enforcement capabilities might change at any time. The result was increased deforestation as villagers exploited the forest in an unrestricted manner.

The Nepalese example provides an interesting contrast between property-rights theory versus practice. The theory assumes clear distinctions between de facto and de jure rights, as well as between the different categories of ownership. The new government policy made continued use of the nationalized land by villagers de jure illegal. From the perspective of the user communities, however, their long historical use of the forest resource gave them a community-based de facto right to continue customary practices, and this right was considered more powerful than any governmental change in national legislation. The government alienated locals and was unable or unwilling, or both, to provide the national forestry agency with the information, personnel, and financial resources required to manage and protect the vast new areas under its jurisdiction. By doing so, the government essentially turned a commons resource into an open-access resource. This decision would eventually prove disastrous for the country's environment and its people—especially in the face of growing population pressures.

In many instances, state or public ownership changes the nature of incentives for surrounding residents to protect forests. Villages, communities, households, and individuals may perceive state ownership as a threat to their traditional user rights and benefits—often with good reason. In turn, state governments often perceive local ownership and management of forests as damaging.

## Traditional Users Versus New Migrants

In addition to conflicts between the state and traditional users, conflicts over user rights exist between traditional users and new migrants. Governments of developing countries have often treated the forest resource as the agricultural frontier, regardless of whether the soil and rainfall characteristics of the area could support either annual or perennial agriculture. In their desire to relieve general population pressures, lower the number of unemployed urban poor, ease political pressure for land redistribution schemes, or provide clear national claims in sparsely inhabited areas, governments have both formally and informally promoted the migration of thousands of colonizers. These programs and policies have often been supported by both multilateral and bilateral aid agencies. If colonization substantially increases the number of people in proximity to protected areas—whether intended or unintended—the impact on biological diversity is usually severe.

The transmigration program in Indonesia provides a compelling example of conflict between traditional forest users and new migrants. Colonization programs in Indonesia have a long and entrenched history. Transmigration from the main island to the outer islands has been part of the country's development policies since 1905, when the country was still under Dutch colonial rule. The resettlement program has had substantial adverse effects on the land and livelihood of both migrants and indigenous groups (Colchester, 1986).

Traditional forms of swidden (formerly called slash-and-burn) agriculture practiced by tribal societies depended on low population densities, and they were constrained by the poor soils of the outer islands as well as by the limited availability of labor in the communities. When this type of agriculture is practiced in the higher-density settlement areas, the result has frequently been felling of greater areas of forest (Donner, 1987). These generally unsuccessful attempts at resettlement came at a cost of tremendous environmental degradation.

The transmigration program in Indonesia did not give sufficient attention to the traditional user rights of the indigenous population, nor did it consult locals about the likely impact the policy would have on customary practices and community-based culture. While this can be blamed in part on decisions made by the colonial government, Indonesia made a clear decision after gaining independence to continue the relocation programs. As a result, inappropriate agricultural practices were instituted widely, causing large-scale environmental degradation. The policy thus ignored local knowledge of ecosystem interactions, environmental externalities, and the likely impact of added population pressure on very fragile ecosystems. The transmigration program also tended to favor migrants over locals, causing distributional equity problems (Colchester, 1986). Though it continues today, the transmigration project has been a social and economic disaster and has had significant adverse impacts on the biodiversity of the outer islands.

Another example of conflict between traditional users and new migrants comes from Amazonian Ecuador, where indigenous communities and colonists came into conflict as a result of policies promoting private land tenure and cattle raising. Although the Ecuadoran government had not instituted any formal colonization program, agrarian reform laws enacted during the early 1960s and in 1973 encouraged the migration of large numbers of people from the Andean highlands into the Amazonian forest. People sought land over which they could claim private ownership. The population growth rate in the Amazon between 1962 and 1974 was triple that of the national rate. During the same period, the government offered credit for cattle operations to encourage new landowners to convert their forests and subsistence plots to pasture. By the time the 1973 agrarian law was passed, any land out of active cultivation or not "improved" for a period of five years was open to be claimed as property. Such "use it or lose it" policies are common in natural-resource policy, and often lead to unsustainable exploitation of the resource in question (MacDonald, 1992).

Amazonian Indians, in an effort to maintain some of their traditional and customary rights to the land and resources, had to take part in cattle ranching and "improve" portions of their uncultivated lands to retain existing tenure. The government more readily gave titles and credit to individual landowners than to communal landowners, which caused disruptions in communal tenure arrangements. Claiming individual parcels within traditional community lands led to disputes about access to hunting lands and patterns of traditional swidden agriculture. Cattle often damaged agricultural crops and interfered with traditional labor requirements for farming, hunting, and fishing (MacDonald, 1992).

The Ecuadoran government failed to consider the historical context for assigning user rights. By promoting private land tenure and cattle ranching, the government forced the Indian population to begin non-forest-based activities if they wished to retain their customary rights. The policies favored colonizers over traditional users and eventually led to a bitter political struggle. In addition, cattle ranchers became more vulnerable to fluctuations in international commodity prices for beef, and private landowners, facing competitive international markets, were less inclined to consider the long-term and external environmental impacts of their activities.

National colonization programs are often developed without consulting the local indigenous communities that will be most strongly affected. As a result, neither colonizers nor indigenous peoples have clearly defined claims to user rights in demarcated areas. If the government institutes some type of land titling scheme, it usually works to the advantage of the colonizers, who have better access to the legal system. In such situations, the impact on biodiversity is usually severe. In Latin America, the problem is exacerbated when land speculators or livestock ranchers begin buying up colonizers' claims once the fragile soils are rendered infertile for agriculture—usually after only three or four years.

## Traditional Users Versus Commercial Enterprises

Reliance on the forest resource as the agricultural frontier is exacerbated by the twin problem of reliance on the forest resource as a source of economic development. In attempts to create strong export markets and develop domestic industrial capacity, developing-country governments lease forested land through concessions to private commercial enterprises. These large commercial operations often threaten the livelihood of traditional users of the forest resource, and because these activities are capital intensive, they are unlikely to generate sufficient countervailing employment opportunities. In addition to devastating biodiversity within the concession area, these commercial operations may drive locals away from their traditional gathering grounds and toward any land outside the concession claim–including protected areas.

Mexico provides an interesting example of the conflict that often arises between traditional users and commercial enterprises. From the 1950s to the early 1980s, logging companies degraded the resources of forest-dwelling communities in several states. The government granted exclusive forest rights to logging companies for 25 or 30 years without any mechanism for monitoring the companies' practices. The companies were expected not only to extract timber but also to replant forest stands. Firms frequently denied the local people access to their forests. As part of their agreement with the state government, companies paid the communities a small "forest-rights rent," but payments were only a fraction of the value of the extracted timber (Bray, 1991).

More specifically, until the past decade, the Maya of Quintana Roo in the Yucatan received little compensation for the exploitation of their forest land (Bray, 1991). In the southern region, the government granted a concession to the timber company Maderas Industrializados de Quintana Roo (MIQRO). During its 30-year concession (beginning in the 1950s), MIQRO made deals with agrarian collectives, or *ejidos*, to harvest mahogany and cedar selectively. However, the *ejidos* received little in return. MIQRO brought in its own workers and shared little of the profits.

In granting these forest concessions, the Mexican government failed to acknowledge that having a successful forest-management plan depended on having the cooperation of the local community. While MIQRO and its small-scale contractors were supposed to reforest and to allow access to the forest by locals for traditional uses, the company essentially ignored those terms of the contract. The government was either unwilling or unable to enforce the terms, and the local community lost out as a result. The government did not give MIQRO any incentive to consider the long-term environmental impacts of its activities, nor did it clearly specify and enforce the user rights contained in the concession agreement. Frustrated by the uncertainty over appropriate and allowable use granted to MIQRO and by the adverse political and economic distributional impacts on local communities, the Maya eventually coalesced to regain control over the production and marketing of the timber resource.

The Quintana Roo example makes clear that the state's decision to assign private rights to logging companies did not preclude or excuse the government from taking steps to ensure that the forest resource would be managed and exploited sustainably. The Maya's political cohesion, rooted in a network of community-based forest use, eventually led to the reversal of the government's policy of favoring MIQRO at the expense of local users. This in turn led to use of the forest resource in a more sustainable manner. Providing such opportunities to local populations may minimize the number of illegal incursions into protected areas and, if a sound management plan is implemented, may even enhance the protection of biodiversity in the worked-over areas.

## Local and State Interests Versus International Stakeholders

Forests provide benefits or values that extend beyond local forest dwellers. As the circle of beneficiaries from tropical forests widens from forest villages, communities, and households to regional communities and, finally, to the global community, the range of perceived values also increases. Tropical forests are valued for utilitarian, aesthetic, moral, and ecological reasons (Botkin and Talbot, 1992). Forest dwellers—villages, communities, and households—often derive the most immediate value by utilizing forests for vital services that sustain livelihoods and satisfy daily needs. Tropical diversity sustains many of the developing world's indigenous populations by providing a storehouse of materials for shelter, tools, fuel, clothing, and food. Population pressure and distributional concerns, however, can push these same individuals into behaviors that threaten the very biodiversity that sustains them.

While this book focuses on biological diversity, tropical forest ecosystems provide other important environmental services. For example, they provide watershed-protection benefits to a larger regional area as well as to local communities. Forested areas protect watersheds and soils by allowing rainwater to infiltrate into groundwater systems, reducing surface runoff, stabilizing soils through the buildup of organic matter, and protecting soils from wind erosion. When forest cover is reduced, the adverse watershed effects are felt by both the forest's immediate residents and by more distant lowland populations. For instance, upland populations in the Central American highlands and the Central Highlands plateau in Ethiopia suffer from degraded lands from soil erosion and water shortages due to reduced groundwater recharge and lower water tables (WRI, 1985). Lowland populations, such as those near the Himalayan mountains of India, can be devastated by intense flooding and sediment deposition (WRI, 1985). At the same time, agriculture throughout the world has been based to a significant extent on the genetic resources of tropical forest plants (Botkin and Talbot, 1992).

Just as the individual forest species are valued for their potential uses and functions, the tropical forest ecosystem as a whole is valued for its role in climate regulation. On a regional level, tropical forests influence rainfall patterns. The forests in the Amazon are estimated to be the source for 50–80 percent of the region's atmospheric moisture, which is recycled as rain every five and one half days (Mabberley, 1992). Given this ecological function, a balance must be maintained between the amount of land cleared for agriculture and the amount left in forests. Below a certain threshold of forest cover, neither the tropical forests nor tropical agriculture would be viable.

On the global level, forests have a role in the regulation of carbon dioxide, one of the gases linked with global warming. Tropical forest vegetation is thought to contain 500 billion metric tons of carbon (Botkin and Talbot, 1992). Carbon is released from tropical forests principally through forest burning. The contribution to greenhouse gases through deforestation in developing nations versus that through fossil-fuel combustion in developed nations, however, has been a contentious issue (Agarwal and Narain, 1991).

User groups often have limited incentive to take the environmental externalities created by the forest resource into consideration, partly because of their inability to adopt a long-term view of the forest resource. As population and poverty pressures mount, more and more marginal groups move onto marginal lands—regardless of whether they are designated protected areas. Without reasonable alternatives, the pressure rises to make illegal incursions into protected areas, and critical environmental externalities are lost. Even if transnational environmental externalities are ignored, customary use rights may no longer be sustainable in either the short run or the long run.

Determining who has legitimate claims on the tropical forest resource at what times and in what places is a politically volatile task. Clearly, the implications for managing the resource change dramatically depending on who is included as a legitimate stakeholder. Stakeholders have widely varying views on both the importance of biological diversity and the compatibility between biodiversity preservation and resource use. In addition, certain uses of the forest require the exclusion of other uses, which in turn implies the exclusion of certain users.

## CASE STUDIES

The common conflicts surrounding user rights have played out in varying ways throughout the tropical developing world. Governments have had mixed results in achieving biodiversity-management objectives both within and outside protected areas. In this section, we present three case studies that explore how the problems of population pressure and equity can be reconciled with preserving biodiversity through creative user-rights arrangements.

### INBio in Costa Rica

Costa Rica's National Institute for Biodiversity (INBio) is a private nonprofit group. Created in 1989 as the result of a presidential planning commission on natural-resource management, INBio seeks to generate new knowledge and to integrate and disseminate knowledge of Costa Rica's natural resources, with the goal of preserving Costa Rica's biodiversity through the sustainable use of its biological resources (R. Gamez, pers. comm., 1992). This goal arose from the commission's assessment that simply establishing protected wildlands was insufficient to conserve Costa Rica's natural resources.

INBio's efforts focus on creating an inventory of all plant and animal species in the country. The institute hopes to do this within 10 years, under an estimated annual budget of about $3 million raised primarily from private foundations and environmental organizations (Joyce, 1991). INBio coordinates a countrywide program of collection, identification, and classification of specimens. It stores the specimens, analyzes and catalogs them using a sophisticated computer system, and makes them available to researchers. INBio has already received some payoff for creating this type of information base—the institute reached agreement with Merck International for a reported $1 million to provide specimens to the pharmaceutical company.

An important element of INBio's strategy is to establish and increase the intellectual, spiritual, and economic value of biodiversity by changing people's perception of the forest's value to include its biodiversity value. INBio's organizers realize that the project's success depends on the support and cooperation of farmers and other people living near areas of natural and biological significance. Accordingly, the institute has consistently sought to incorporate local participation in their programs (R. Gamez, pers. comm., 1992). It trains farmers and other rural workers to help find, sort, and catalog specimens from the country's national parks. INBio currently has a staff of nine entomologists, four botanists, and 55 "parataxonomists." The parataxonomists are farmers, housewives, bus drivers, university students, former park guards, and other laypeople; INBio has specifically concentrated on recruiting women and individuals with lower incomes into its workforce.

The institute provides local people being trained as parataxonomists with a six-month training session that includes lessons on botany, entomology, and ecology, and also regularly conducts refresher courses for the staff. The parataxonomists are sent to various biological reserves and parks around the country, and workers split their time between their homes and field stations in the parks. The parataxonomists collect specimens of about 5,000 insects and 50 plants per month. They are also encouraged to act as "tropical forestry educators" in their home communities (R. Gamez, pers. comm., 1992).

INBio maintains a cooperative relationship with the government, in particular with the Ministry of Natural Resources (MIRENEM). Although INBio gives MIRENEM a portion of the income it receives (from providing specimens to pharmaceutical companies and other interested organizations) for the management of national parks, the institute has no say regarding which parks the funding supports. INBio's director would prefer that the money be channeled to the particular parks from which the specimens were obtained. The idea is that this would provide local people with increased incentive to conserve the biological resources in their region and to use them to compete for research and other opportunities (R. Gamez, pers. comm.).

Through both direct employment and educational outreach efforts, INBio has helped foster a desire among local populations to maintain protected forest areas. The institute's approach explicitly acknowledges that successful forest management depends on the cooperation of the local community. The approach holds great potential for promoting long-term and sustainable management of the forest resource.

Several factors account for the relative success of the Costa Rican project:

- The Costa Rican government made a clear philosophical commitment to the national park system. This commitment was then institutionalized and protected somewhat from the political process through the creation of several autonomous nongovernmental organizations (NGOs), including INBio.
- Creative leaders established clear links between the protection and preservation of biodiversity and the economic prosperity of the country. By capitalizing on international interest in both ecotourism and pharmaceutical prospecting, biodiversity became a reliable source of foreign exchange.
- INBio established professional and financial relationships with international NGOs. In doing so, it created an alternative funding source from the government and was able to take advantage of the NGOs' scientific and administrative expertise. In addition to giving INBio international media exposure, these connections freed the project from some fluctuations in funding and personnel that are a natural outgrowth of the political process.
- Through the parataxonomists program, INBio vested locals in the maintenance of biodiversity. In addition to direct employment, the program provided critical educational services to the wider population. Through its emphasis on training women and lower-income individuals, the program directly addresses equity considerations.

INBio has been successful in advancing its goal of inventorying the biodiversity of Costa Rica. It is important to remember, however, that this success has occurred in a country where the vast majority of both the urban and rural populations have access to potable drinking water, health services, and electricity. In addition, the population growth rate has slowed substantially in the past decade. Even so, this example seems to illustrate that true state-owned protected areas with no direct user rights for local populations, if combined with appropriate educational and employment opportunities, can succeed in protecting biological diversity.

By vesting the local population in protecting the area's biodiversity, INBio has eased some of the pressure on the national park system and has encouraged Costa Ricans to consider explicitly and systematically the environmental impact of their forest practices. Through its training programs, INBio fosters a more equitable return to investing in forest resources. If MIRENEM did channel at least part of its revenues to the parks from which specimens are obtained, the outcome might be an even better incentive system and use arrangements that are more equitable.

## Chimane Forest Reserve in Bolivia

An interesting example of how changes in user rights affect biodiversity is provided by the first debt-for-nature swap, signed in July 1987 between Bolivia and Conservation International, a private U.S. environmental organization. Under the agreement, Conservation International purchased $650,000 in Bolivian debt from a Swiss bank for about $100,000, a discount of almost 85 percent from the nominal value of the debt. In return for being able to exchange the debt for domestic currency obligations, the Bolivian government agreed to extend an existing national park in the Beni region of the Amazon to an area more than 10 times its original size and to bestow upon the expanded area of the Beni Biosphere Reserve the highest degree of legal protection. Conservation International offered technical, scientific, and administrative assistance to implement resource-management programs for the park and for adjacent protected development, watershed, and resource zones. To provide for the long-term success of the swap, the Bolivian government, with assistance from the U.S. Agency for International Development, agreed to establish an endowment fund of $250,000 in local currency to protect and maintain the Beni park (Page, 1989; Winthrop, 1989).

The program, however, faced problems in the field. In the "permanent production" zones, the Chimanes, Mojeno, Yuracare, and Movima tribes complained that the area had been subjected to further degradation as a result of the debt-for-nature swap. The Indians claimed that allowing logging companies into the area for "sustainable forestry" had resulted in overexploitation of the forest and the loss of local subsistence, and had contributed to an influx of colonists traveling on logging roads (Painter, 1990). The Indians maintained that the implementation of the debt-for-nature swap had prevented them from gaining title to their lands and from protecting their homes. In addition, the Indians were upset that neither the government nor Conservation International had consulted them before approving logging activities.

In August 1990, the indigenous Indians organized a march to the Bolivian capital to demonstrate their unhappiness with the logging companies and to demand that the government recognize the land in the Chimane Forest as Indian territory. The Indians won out over a coalition that included the logging companies, Conservation International, and the International Tropical Timber Organization, and they gained control of 1.9 million acres of land in the Beni tropical lowlands (*Boston Globe*, 1990). A new law enacted in September 1990 provided for the gradual phase-out of logging operations in the area in preparation for ceding the lands to the Indians.

The action by the Indians was the first-ever centralized organization of the seminomadic tribes. The activists stressed that their primary concern was gaining access not to the timber but to the land. The Indians also dismissed the idea that they planned to convert the land for agricultural production. They pointed out, "We have lived in these woods since our ancestors. We have learned to care for and maintain its ecology because we know it guarantees our existence" (Collett, 1989:18). The Indians planned to own and manage their land as a "common territory" in order to preserve it as an important ecosystem. In addition, the Indians volunteered to serve as forest wardens.

The case of the Chimane Forest Reserve is very complicated, and the implications for the protection of biodiversity are much less clear. The mixed results of the project can be accounted for by several factors:

- The project did manage to secure a regular source of funding through the establishment of an endowment generating income for annual operating expenses. This should help to shelter the project from the ups and downs typical of budget cycles.
- The decision to transfer 1.9 million acres of land to the indigenous Indians did much to advance equity concerns in regard to colonizers and commercial enterprises. The extent to which the outcomes are truly equitable, however, will depend on how the Indians themselves choose to assign user rights across gender and ethnic lines. In addition, the Bolivian government did not secure any guarantees that biodiversity preservation would remain a management priority for the Indians.
- By establishing "permanent production zones" in areas immediately adjacent to the Beni Park, the project created income opportunities without providing mechanisms for limiting the subsequent influx of colonizers into the area. Without such limitations, in-migration tends to overwhelm the local infrastructure and pushes both locals and migrants into harvesting within the protected areas. The permanent production zones may have been more appropriately located at some distance from the park.
- The project failed to recognize that locals and migrants might consider the permanent production zones as additional rather than substitute resources to the park. Especially given the poverty that characterizes the area, project leaders should not have been surprised that people aggressively sought a higher standard of living than the permanent production zones were able to provide.

In many ways, the Bolivian example is one where the project was trying too hard to be all things to all people—including conservationists, loggers, colonizers, and indigenous Indian groups. The project may have performed better if specific management priorities were explicitly identified with specific geographic areas, that is, stating that in this area, biodiversity preservation will take priority, whereas in this other area, empowering local Indian groups will take priority. Even with this type of prioritization scheme, however, locals still need to be vested, through education or income opportunities, in the preservation of biodiversity.

Unwilling to accept the vast distributional and environmental impacts of the original swap, the Bolivian Indians coalesced to renegotiate the terms of the agreement. This led to explicit recognition of community-based ownership patterns and more sustainable use of the forest resource. If spillovers from environmental externalities are contained within the group-ownership area, communal property schemes may better take into consideration the ecological impact and long-term sustainability of management practices. In the face of growing population pressure, however, it is unclear how often the Bolivian government (or any government) would be willing to turn over such a large expanse of land to such a limited population—even if the low population density is better suited to the regional ecosystem.

## Community Baboon Sanctuary in Belize

In many cases, ecotourism is encouraged by the governments of developing countries. The governments may seek to promote their country as a haven of natural beauty in order to attract international tourists who provide foreign exchange to the local economy. This may be done by designating areas of the country as national parks or protected areas, by assisting in the development of tourist enterprises such as lodging, and by conducting advertising campaigns abroad to attract visitors. Since ecotourism confers benefits upon the local inhabitants of an area, however, it has also arisen without government intervention. One such case is the Community Baboon Sanctuary in Belize, a heavily forested country in Central America (Horwich and Lyon, 1990; Lipske, 1992).

In 1985, in the forest village of Bermudian Landing, a small group of local landowners was approached by a U.S. scientist doing research in the village. Robert Horwich was searching for a way to preserve the habitat of the black howler monkey (known locally as "baboons"), which lives in the forests around the village. At issue was the use of land in the community. The local inhabitants typically practiced swidden agriculture. Tracts of land were burned to create small fields, or *milpas*. The *milpas* were farmed for a few years and then were abandoned to regenerate into young forests while the villagers moved on to clear another plot.

Horwich presented a plan in which local landowners would preserve skeletal tracts of forest land alongside agricultural fields and river banks. While the tracts would be small enough that agricultural production would not be significantly reduced, the area preserved would be sufficient for the howler monkeys to exist by traversing between *milpas*, uncut mature jungle, and recently rejuvenated forest. In addition, the local community would benefit from reductions in soil erosion and siltation, from maintenance of the fish stock in the nearby river, and from more rapid regeneration of the *milpas* after farming.

The plan was designed to thrive entirely on voluntary participation by the local people. It was approved by the government of Belize and funded by the World Wildlife Fund, a U.S. environmental organization. In 1985, with the agreement of the village council and several local farmers, a preliminary site was selected and then examined and documented according to its existing vegetation, property lines, and its use by howler monkeys. The mapping was used to design management schemes for individual plots. The landowners then voluntarily signed pledges stating they would maintain their plots in order to protect the habitat of the monkeys.

After being persuaded of the value of environmental conservation and the preservation of the howler monkey, the villagers perceived that the presence of the noisy primates might be of interest to others as well. The local people suggested that tourists, who might initially be attracted to Belize by the tropical coastal reefs, could also visit their community. In this way, participants in the project would be able to supplement their agricultural incomes with tourist revenues.

Thus, to the original goals of conservation were added the goals of education, tourism, and research. In 1987, a natural history museum was constructed in Bermudian Landing. Belize's first museum, it contains exhibits on the area's natural resources, conservation, and local culture. Visitors can tour the system of interpretive trails and purchase educational materials published by the sanctuary. Beyond the sanctuary, its manager conducts lecture tours in both rural and urban schools in Belize. And recently the area has come to be used for research on topics such as the howler monkey, riverine forest, and Creole culture.

Today, tourism is a vital industry in the area. In 1992, more than 6,000 visitors came from around the globe to glimpse the howler monkeys. In addition to providing guides for the forest area, the villagers offer room and board in their own homes. This entrepreneurship is being aided by a $10,000 grant from the Inter-American Foundation. The grant is to be disbursed by the Belize Audubon Society as revolving loans to villagers constructing lodging or other tourist facilities.

Since its beginning with 11 villagers and a three-mile expanse along the Belize River, the project has expanded to include eight villages, more than 100 landowners, and an area of about 18 square miles. It is estimated that approximately 90 percent of the landowners are keeping their commitments, despite the fact that management plans are not legally binding. Consultations between the landowners and sanctuary managers, and educational programs for the participants, encourage adherence to management plans. In the future, participants' performance will be monitored with aerial photography.

The government facilitated development of the plan by recognizing private ownership rights, while international and local NGOs provided financial support and technical assistance. The crucial motivations for the sanctuary's development, however, rested in the fact that its implementation provided clear and regular benefits, integrated conservation with other economic activities, and mobilized local exploiters to maintain the resource. Strong community groups that became involved in a variety of managerial, educational, and technical tasks were ideal sites for the dissemination of information and rewards. The ability of these groups to operate successfully was aided by the organizational discipline and autonomy of NGOs associated with the project, such as Community Conservation Consultants and the Community Baboon Sanctuary groups.

This case study demonstrates how purely private property rights—that is, full devolution to the local population—can be made consistent with the preservation of biodiversity preservation outside protected areas. Its success depended in part on these elements:

- The project designers understood that assigning user rights has inherent economic and political distributional components. Rather than asking landowners to give up important sources of income, the project created an environment in which what the farmers did in their own best interest was consistent with biodiversity preservation.
- The project provided technical expertise and had an important educational component. Thus, in addition to financially vesting the farmers in the preservation of black howler monkeys, the project also encouraged a philosophical and intellectual commitment to biodiversity.
- The project took advantage of linkages to an international nongovernmental organization and linkages to international ecotourism markets. In addition, the government played a supportive role by approving the area's status as a sanctuary.
- The project was designed to remain financially stable. While the Inter-American Foundation provided some seed money, the project depends for the most part on internal resources.
- The project respected—and in fact, depended on—the entrepreneurial capabilities of the local populations. The expansion of the preservation project into an income-generating, environmentally sensitive ecotourism project resulted directly from creative leadership within the community.

The farmers associated with the Community Baboon Sanctuary did much more than might otherwise be expected to preserve and protect biodiversity. Owners of private property may be poorly motivated to consider the large-scale ecological impacts of activities undertaken on their private land. They may, however, be quite interested in preserving environmental values for their own consumption and their own income-enhancing opportunities. Such was the case in Belize. In addition, developing this type of preservation ethic on private lands may help the cause of the true protected areas.

The Community Baboon Sanctuary in Belize presents a successful model for providing alternative sources of revenue through ecotourism. Through this project, the community was able to reconcile broader societal interests in biodiversity with local needs for agricultural production; obtain technical assistance regarding what was best for the howler monkey population and how this might impact short-term and long-term agricultural productivity; create an international market for local biodiversity; tap into marketing efforts for tropical-reef tourism at the national level; and create income opportunities for both individuals and communities.

## DISCERNING AN APPROPRIATE PATH FOR THE FUTURE

The appropriateness of devolving user rights to local populations will depend greatly on the social, ecological, and economic context for the decisions. The case studies highlight three alternative situations: the INBio project in Costa Rica revolves around true protected areas; the Chimane Forest Reserve in Bolivia combines true protected areas with community-managed buffer zones and extractive reserves; and the Community Baboon Sanctuary in Belize operates within private landholdings. With varying success, each project has addressed directly or indirectly the protection of biological diversity.

By establishing true protected areas, governments explicitly exclude the possibility of devolving user rights to local populations. The state may or may not have sufficient financial and personnel resources to enforce such use rules or to implement management plans for the protected areas. In developing countries, constraints on infrastructure and enforcement capabilities often prevent successful management (Feeny et al., 1990). Even if the government has sufficient resources to enforce use restrictions fully, it still may not implement sustainable management practices in the protected areas. Political favoritism, short time horizons, or a limited understanding of the ecosystem may all lead to inappropriate policies.

Maintaining forested protected areas under public ownership can contribute substantially to the more appropriate use and preservation of the forests and the natural products and environmental services that they provide. Government ownership may sometimes be the most effective way to consolidate natural areas, to promote awareness of and respect for the resources, and to conserve the resource. Government ownership is commonly promoted to provide for pure public goods and benefits that are not captured by markets. When forest lands are held under public control, however, it is important to design conservation programs so that the local people are involved in the implementation of preservation efforts and are motivated to participate in the maintenance of the forest.

Owners of private property usually have great incentives (e.g., market prices, accrual of benefits and costs, rights of transfer) to regulate resource use. In developing countries with agriculturally based economies, however, poverty is crucially linked to degradation. When market prices are exogenous and do not reflect local scarcity, intensified agricultural production to meet subsistence needs can lead to resource degradation (Larson and Bromley, 1990). In addition, the economic incentives do not usually extend to consideration of environmental externalities, and private owners may thus overexploit the resource from a socially optimal point of view, even under the best macroeconomic conditions. Governments must then struggle to develop an incentive system that encourages wise investments by private landowners with respect to ecological considerations.

In deciding whether to devolve user rights, it is important to consider the historical context for assigning property rights. If there is still a strong tradition of communally based ownership or customary use rights, project designers should take advantage of these relationships. But if these relationships have long been destroyed, it may be pointless to try to recreate ownership patterns that are wholly foreign to the current generations. Private or state-held lands can both be made to work toward the protection of biodiversity.

Clearly, however, open-access arrangements of property rights are inappropriate. This policy is especially problematic in areas experiencing significant population pressure. Government-owned land must be accompanied by the provision of sufficient personnel and financial resources to ensure their protection. Without this, state-held reserves essentially turn into open-access resources. Even without providing local people with user rights, the locals can be vested in publicly held lands by creating employment opportunities or by creating an environmental ethic. In both cases, training and education will be critical to the success of any management approach.

Individuals, households, communities, and governments must tailor user rights to fit the country's social, political, economic, and ecological characteristics. Carefully considered user-rights policy has the potential to create an environment in which all forest users are engaged in sustainable activities. Ideally, what user groups want to do will be consistent with what planners would optimally have them do, regardless of how any external factors change over time. The most appropriate way to accomplish this is to vest resource users in the long-term maintenance of the forest resource. While programs addressing health and social problems may also be needed, community initiation and management of forest areas can improve self-sufficiency and self-reliance and can address poverty by extending new technical skills, markets, and opportunities to depressed rural areas.

This chapter has discussed a variety of arrangements in its exploration of the impact of devolution decisions on biodiversity protection. The case studies demonstrate that no one arrangement represents the "best" option. In fact, it may be more appropriate to consider a portfolio of user-rights arrangements that vary across the landscape. The desirability of devolving user rights to a particular piece of land will depend on the importance placed on various land-management objectives. Explicitly acknowledging that biodiversity will take precedence in some areas (presumably the protected areas) but will be secondary in other areas (e.g., agricultural settlements) may better serve all stakeholders.

International cooperation regarding forest preservation for the sake of interests beyond the local or regional community may be a much-needed effort. For developed nations to avoid accusations of environmental imperialism, they must acknowledge that the industrialized world has relied on massive exploitation of their own forests and the forests of developing countries for centuries, while environmental awareness is only relatively recent and still only gradually being incorporated into land-management policies. The consequences of forest loss have been seen and repeated throughout developed and developing countries, and the responsibility for preserving global environmental quality is global.

## NOTES

Thanks to Randall Kramer and Carel van Schaik for their insightful substantive advice as well as their editorial suggestions. We are also grateful to William Ascher, Clark Binkley, Gardner Brown, Lynton Caldwell, Karlyn Eckman, Hans Gregersen, William Hyde, Owen Lynch, and Narendra Sharma for comments and criticisms on early versions of the manuscript. Able research assistance was provided by Catherine Karr and Christopher Jones.

## REFERENCES

Adeyoju, S. Kolade. 1976. Land use and tenure in the tropics. *Unasylva* 28:26–47.

Agarwal, Anil, and Sunita Narain. 1991. Technology control, global warming and environmental colonialism: the WRI report. *Social Action* (Jan.–Mar.): 3–28.

Aufderheide, Pat, and Bruce Rich. 1988. Environmental reform and multilateral banks. *World Policy Journal* (Spring): 301–21.

*Boston Globe*. 1990. Bolivia cedes forest to Indians. 26 September, p. 2.

Botkin, Daniel B., and Lee M. Talbot. 1992. Biological diversity and forests. In *Managing the world's forests*, Narendra Sharma (ed.), 47–74. Dubuque, Iowa: Kendall/Hunt.

Bray, David Barton. 1991. The struggle for the forest: conservation and development in the Sierra Juarez. *Grassroots Development* 15 (3): 13–25.

Bruce, John W., and Louise Fortmann. 1988. Why land tenure and tree tenure matter: some fuel for thought. In *Whose trees: proprietary dimensions of forestry*, Louise Fortmann and John W. Bruce (eds.), 1–9. Boulder, Colo.: Westview Press.

Colchester, Marcus. 1986. Unity and diversity: Indonesian policy towards tribal peoples. *The Ecologist* 16:89–98.

Collett, Merrill. 1989. Debt deal stacked against Indians. *The Progressive* 53 (8): 17–18.

Commander, Simon. 1986. Managing Indian forests: a case for the reform of property rights. *Development Policy Review* 4:325–44.

de Saussay, Chris. 1987. *Land tenure systems and forest policy*. Bognor: U.N. Food and Agriculture Organization.

Donner, Wolf. 1987. *Land use and the environment in Indonesia*. Honolulu: University of Hawaii Press.

Ecumenical Center for Documentation and Information. 1989. Indian rights in the new Brazilian constitution. *Cultural Survival Quarterly* 13 (1): 6–12.

Feeny, David, Fikret Berkes, Bonnie J. McCay, and James Acheson. 1990. The tragedy of the commons: twenty-two years later. *Human Ecology* 18 (1): 1–19.

Gibbs, Christopher J., and Daniel Bromley. 1989. Institutional arrangements for management of rural resources: common property regimes. In *Common property resources: ecology and community-based sustainable development*, Fikret Berkes (ed.), 22–32. New Brunswick, N.J.: Belhaven Press.

Gregersen, Hans, Sydney Draper, and Dieter Elz. 1989. *People and trees: the role of social forestry in sustainable development*. Washington, D.C.: World Bank.

Guha, Ramachandra. 1983a. Forestry in British and post-British India: a historical analysis. *Economic and Political Weekly* 18 (43): 1882–96.

Guha, Ramachandra. 1983b. Forestry in British and post-British India: a historical analysis. *Economic and Political Weekly* 18 (44): 1940–46.

Hardin, Garrett. 1968. The tragedy of the commons. *Science* 162:1243–48.

Horwich, Robert H., and Jonathan Lyon. 1990. *A Belizean rain forest: the community baboon sanctuary*. Gay Mills, Wis.: Orang-utan Press.

Jacobson, Jodi L. 1992. Out of the woods. *World Watch* 5 (6): 26–31.

Joyce, Christopher. 1991. Prospectors for tropical medicines. *New Scientist*. 19 October, 132 (1791): 36.

Larson, Bruce A., and Daniel Bromley. 1990. Property rights, externalities, and resource degradation: locating the tragedy. *Journal of Development Economics* 33:235–62.

Leonard, H. Jeffrey. 1987. *Natural resources and economic development in Central America: a regional environmental profile*. Published for the International Institute for Environment and Development. New Brunswick, N.J.: Transaction Books.

Lipske, Michael. 1992. How a monkey saved the jungle. *International Wildlife* 22 (1): 38–42.

Mabberley, D. J. 1992. *Tropical rain forest ecology*. London: Blackie.

MacDonald, Theodore. 1992. From reaction to planning: an indigenous response to deforestation and cattle raising. In *Development or destruction: the conversion of tropical forest to pasture in Latin America*, Theodore Downing, Susanna B. Hecht, Henry Pearson, and Carmen Garcia-Downing (eds.), 213–34. Boulder, Colo.: Westview Press.

Menzies, Nicholas. 1988. A survey of customary law and control over trees and wildlands in China. In *Whose trees: proprietary dimensions of forestry*, Louise Fortmann and John Bruce, 51–62. Boulder, Colo.: Westview Press.

Millon, Rene F. 1955. Trade, tree cultivation, and the development of private property in land. *American Anthropologist* 57 (4): 698–712.

Obi, S. N. Chinwuba. 1963. *The Ibo law of property*. London: Butterworths. Excerpted in Louise Fortmann and John W. Bruce (eds.), *Whose trees: proprietary dimensions of forestry* (Boulder, Colo.: Westview Press, 1988), 34–39.

Page, Diana. 1989. Debt-for-nature swaps: experience gained, lessons learned. *International Environmental Affairs* 1 (4): 275–88.

Painter, James. 1990. Bolivian Indians protest wrecking of rain forest. *Christian Science Monitor*. 18 September, p. 4.

Poore, Duncan, Peter Burgess, John Palmer, Simon Rietbergen, and Timothy Synnott. 1989. *No timber without trees: sustainability in the tropical forest*. London: Earthscan Publications.

Schlager, Edella, and Elinor Ostrom. 1992. Property-rights regimes and natural resources. *Land Economics* 68 (3): 249–62.

Thiesenhusen, William C. 1991. Implication of the rural land tenure systems for the environmental debate: three scenarios. *Journal of Developing Areas* 26:1–23.

Troup, R. S. 1940. *Colonial forest administration*. New York: Oxford University Press. Excerpted in Louise Fortmann and John W. Bruce (eds.), *Whose trees: proprietary dimensions of forestry* (Boulder, Colo.: Westview Press, 1988), 280–83.

Winthrop, Steven vanR. 1989. Debt-for-nature swaps: debt relief and biosphere preservation? *SAIS Review* (Summer/Fall): 129–49.

WRI [World Resources Institute]. 1985. *Tropical forests: a call for action*. Washington, D.C.: WRI.

WRI [World Resources Institute]. 1992. *World resources: 1992-1993*. New York: Oxford University Press.

WRI [World Resources Institute]. 1994. *World resources: 1993-1994*. New York: Oxford University Press.

# 8

# Tropical Forest Biodiversity Protection: Who Pays and Why

*Randall A. Kramer and Narendra Sharma*

People value biodiversity found in tropical rain forests for a variety of utilitarian, aesthetic, moral, ecological, and socioeconomic reasons (Botkin and Talbot, 1992). For instance, traditional medicines derived from plant and animal species found in the tropics provide health services to rural and urban populations; about 25 percent of the pharmaceutical products produced in the United States are associated with plants (WRI et al., 1992). Genetic materials extracted from plant and animal species have contributed to the development of commercial agricultural products (e.g., new varieties of wheat, maize, and rice) that are more resistant to pests and diseases. And nature tourism, often associated with protected wildlife habitats, has become an important source of income, generating about $12 billion annually in worldwide earnings (Lindberg, 1991).

There are important socioeconomic and political considerations in the valuation of biological resources and the protection of biodiversity. First, the benefits that result from biodiversity have spatial and temporal dimensions. The ecological services linked with biodiversity, such as clean air and water, and the use of genetic material and ingredients extracted from plants, animals, and microorganisms, occur at different places and at different times, often beyond the "economic time scale" of individuals.

Second, biodiversity has characteristics of a public good locally and nationally and may be considered a "global environmental good" in an international context. The benefits of public goods flow to all people regardless of whether they have paid for the good, which means that public goods suffer from the problem of "free riders." In a national context, economists have long focused attention on the difficulty of financing public goods and have generally concluded that such goods will be underprovided by markets. In the international context, the provision and financing of public goods is even more problematic. These characteristics make management of biodiversity institutionally complex and create problems in defining property rights.

Third, conservation of biodiversity can create significant nonuse values. By its very existence, biodiversity can generate economic value without requiring actual use and can provide value by leaving open the option of future use. Environmental economists refer to the first case as "existence value" and the second as "option value." Alternatively, both types of economic value are sometimes referred to as "passive-use value."

Fourth, the world community has limited knowledge about the dynamics of tropical forest ecosystems and the role they play in maintaining natural systems and supporting human well-being. Hence, a risk-averse strategy may suggest erring on the side of caution until the scientific uncertainty is reduced.

Fifth, there are diverse concerns and interests at the local, national, and international levels. The stakeholders at each level interact with and benefit from tropical forests in different ways. Local people, for example, depend on forests for their basic needs and livelihoods—the forests generate food, fiber, fodder, small timber, and fuelwood. At the national level, governments often see forests as an important source of foreign exchange earnings, revenues, employment, and land for agricultural expansion. By contrast, at the international level, the world community views tropical forests as a storehouse for biodiversity and a counterbalance to an increasing greenhouse effect. Consequently, biodiversity protection has different implications for various stakeholders, and each management option may imply a different distribution of gains and losses.

These factors combine to create a dilemma for individual nations as well as the world community in dealing with the biodiversity imperative on a significant scale. In this connection, there are a number of questions that need to be addressed. What is the value of biodiversity, and what are its key determinants? How should this value be determined, especially in economic terms? To what extent can the market reflect the true value of biodiversity? What are the direct and indirect costs of protecting biodiversity? And who should pay for conservation of biodiversity if benefits accrue to the world community and to present and future generations?

The scope of this chapter is limited to the question of costs and benefits associated with protection of tropical biodiversity and how these costs and benefits should be shared among various stakeholders reflecting local, national, and global interests. While the emphasis is on protection of biodiversity in parks and reserves, we briefly examine the wider context of social, economic, and political factors that influence conservation and development of tropical forests throughout the landscape, not just in protected areas. We also examine the valuation of benefits and costs relating to biodiversity, including techniques of measurement, and we follow with a discussion of national environmental action plans, international funds for biodiversity conservation, and the use of trust funds. Finally, we examine the tradeoffs in designating lands for biodiversity protection, and we outline an agenda for action, including research priorities.

## POLICIES, MARKETS, AND MARKET FAILURES

It has become increasingly apparent that there are competing claims to tropical forests, resulting from diverse land-use activities. Economic activities such as agriculture, cattle ranching, fuelwood gathering, commercial logging, mining, and infrastructure development are direct causes of tropical deforestation and may also put pressure on protected areas. But these alternative land uses or causal factors are driven by economic, social, and political forces in a broader context. These forces manifest themselves through market and policy failures, population pressures, poverty, and international trade. The relative importance of these direct and underlying causes of deforestation, and their effects on protected areas, varies significantly among tropical countries.

Government incentive policies drive many of the economic decisions about the use of forest resources. These include concession policies that guide timber harvest on public lands, pricing policies and royalty systems for stumpage and fuelwood, and wood-industries policies. Concession policies (royalties, licensing systems, allowable harvesting methods, amount of allowable harvest, and length of concession period) have been criticized for frequently failing to provide incentives for sustainable management of forests and for reforestation. In countries that have a deficit of wood, pricing policies covering fuelwood and charcoal may undermine conservation and discourage tree planting and more efficient use of fuelwood, thus accelerating depletion of forests and woodlands and contributing to land degradation and biodiversity losses. In the wood-processing industry (sawmills, plywood mills, and pulp and paper mills), credit subsidies, tax provisions, trade barriers against imported wood products, and bans on the export of logs have accelerated investments in some countries and led to the development of inefficient industries and to deforestation (Vincent and Binkley, 1992). While these forest policies are targeted at lands outside protected areas, they may also be a determinant of encroachment pressure on protected areas as other lands available to meet production needs are mismanaged. In addition, policies operating in the broader landscape may affect biodiversity in protected areas, because fauna frequently move between protected and unprotected areas.

Deforestation is often induced by nonforest policies as well. These include policies related to agriculture and livestock, settlement, credit, land use, and taxes (Rowe et al., 1992), which generally provide disincentives for efficient management of forest lands and encourage conversion to other land uses. In Brazil, for example, government-subsidized cattle ranching has been a leading contributor to transformation of forest lands (Binswanger, 1989). Nonforest policies also affect the sustainability of land use in buffer zones around protected areas.

Institutional arrangements, including rules and regulations, also affect forestry. In particular, forest institutions, land tenure, and property rights influence how forests are used and managed. The forest institutions in most developing countries are mainly administrative organizations, with limited capacity and resources for supporting multiple objectives related to conservation and development. Often, the institutions perform poorly in long-term investment planning, preparation, and implementation of forestry projects; enforcement of existing rules and regulations; and monitoring of the forestry situation. This problem is frequently compounded by the fact that governments place a low priority on the forest sector and look at forests as a resource to be mined. Another way in which institutional arrangements affect forest use is that governments, which often hold forest property rights, have undermined the traditional rights of local communities and tribal groups to make use of forests. This has made forests more vulnerable to problems of open access (see chapter 7).

Governments in tropical countries are often reluctant to devote large amounts of land and financial resources to protected areas because the areas do not provide the revenues or taxes associated with using forests for timber extraction. With the exception of tourism, most of the benefits arising from protection do not result in cash flows that governments or conservationists can point to as justifying the protection activities; the benefits of protection are usually less tangible than a truckload of logs or rattan. For example, one set of outputs from the forests—environmental services such as climate regulation or watershed protection—are nonmarket public goods that provide benefits at the regional or national level. Even in countries where these environmental services are recognized as important, their economic values are generally not measured or given the same importance as timber revenues. Nevertheless, as improved ecological and economic information becomes available about these benefits, governments may be more willing to devote resources to protected areas to generate environmental benefits for their citizens.

Protection of tropical forests also produces another flow of benefits that are global in nature and therefore more difficult to justify from the standpoint of within-country benefits. Increasing concerns in developed countries about the role that tropical forests play in carbon cycles and about the conservation of genetic resources produce a class of beneficiaries who live thousands of miles from the locales where protection activities take place. As a result, many of the benefits of protection efforts accrue outside the country where the protection costs are incurred. While some of these benefits derive from future pharmaceuticals and other products developed from protected species, other benefits are more intrinsic in nature (van Schaik et al., 1992). Many people value tropical forests and the biodiversity they contain, even if they have no plans to make direct use of the forests or their products. Economists point to contributions to organizations such as World Wildlife Fund as evidence of these values. Donations to conservation organizations, however, may be significantly smaller than the true level of nonuse values due to free-riding behavior. Thus, they should be viewed as lower bounds (Freeman, 1993). Such contributions may, of course, also reflect use motivations, to the extent that the organizations provide information and other private goods (e.g., magazines, t-shirts, bumper stickers) to their members.

The public-goods nature of some forest outputs makes it unlikely that forest management driven entirely by market signals will result in the mix of landscapes, habitats, biodiversity, and other goods and services that are socially optimal. Hence, some government intervention will be required. One way to ensure that at least representative samples of biodiversity are preserved and to ensure that natural areas are available for recreation by future generations is to establish publicly owned parks. Because parks and reserves do not generate the magnitude of cash flows that timber or agriculture do, innovative means of financing will be necessary, including financing mechanisms that involve the international movement of funds from beneficiaries located far removed from the parks and reserves where protection activities take place.

## MEASURING THE BENEFITS AND COSTS OF BIODIVERSITY PROTECTION

While arguments in favor of protecting biodiversity are often made on ecological, ethical, and political grounds, policymakers often ask about the economic consequences of alternative policy or management options. The economic answers given are too often based on incomplete analysis, focusing on the more obvious and easy to measure costs and benefits associated with marketed products. Applying a more complete accounting framework to biodiversity questions is not easy, but that is not a compelling reason to avoid such analysis. Failure to measure or to fully measure the benefits of biodiversity protection may lead policymakers to implicit assignment of low or zero benefits. Similarly, failure to measure the full costs of biodiversity protection may lead to policy decisions that are unworkable or unwise. Here, we review the use of economic-valuation techniques to measure the economic impacts of protecting or losing biodiversity.

### Economic-Valuation Techniques

Environmental economics emphasizes human preferences in assigning values to nonmarket goods and services such as forest biodiversity. This is admittedly an anthropocentric approach, but most public decisions favor actions that are designed to improve human welfare. (Human welfare can, of course, include an altruistic concern for the survival and welfare of other species.) As a result, the environmental economic approach to valuing biodiversity focuses on individuals' willingness to pay (WTP) for use and nonuse attributes of biodiversity (Randall, 1986). According to Smith (1990), this approach reflects a fundamental change in what is generally meant by consumption.

The point of measuring environmental values is to be able to put them into a benefit-cost framework to compare them with more conventional benefit and cost streams. Measuring WTP for biodiversity protection is challenging because of its many attributes. To the extent that biodiversity is a potential source of future products, its value would be the sum of the future value of all products developed from a particular storehouse of genetic resources and other products that can be used for human consumption. But scientists are unable to predict the probability of developing a successful product from a particular collection of genetic resources, which makes economic valuation of consumptive use problematic (Principe, 1991; Sedjo, 1992).

Another source of economic value from protecting biodiversity in the wild is ecotourism. This value can be estimated using the travel-cost method and other types of recreation-demand models (Kramer et al., 1992). Little is known about the role of biodiversity in tourism, but conceptually it is possible to separate and value the contribution of biodiversity to the overall value of a natural area to tourists. Empirically, this might be done by applying one or more recreation-demand models (for a review of these models, see Bockstael et al., 1991). In general, the models proceed by imputing a value to a particular site by measuring the total expenditures of visitors in terms of time, transportation costs, and out-of-pocket expenses. (See Box 8-1 for an example of the use of this method to value a park in Costa Rica.)

To properly value biodiversity, it is necessary to use a model that allows a separation of the different components of the total recreational-use value of a park. Since there are many characteristics that contribute to the attractiveness of a particular park or natural area, using a simple travel-cost model, which does not allow such separation, can overvalue biodiversity. One way to separate the contribution of biodiversity to the value of a park would be to use the hedonic travel-cost method, which relates tourism expenditures to site characteristics (Brown and Mendelsohn, 1984). This model has been used in the United States to value characteristics of water quality and fishing success, and it could be adopted to compare the effect of differing degrees of perceived biodiversity on tourism activities.

An alternative approach would be to use what is known as a random-utility model. By collecting site-visitation data and data on the characteristics of a number of substitute recreational sites, one can use statistical analysis to examine how people choose among alternative destinations.

*Box 8-1* Using the travel-cost method to value nature tourism in a tropical rain forest

What is the recreational value of a tropical rain forest reserve? This is the question posed by Tobias and Mendelsohn (1991) in their study of the Monteverde Cloud Forest Biological Reserve, in Costa Rica. This 10,000-hectare reserve along the continental divide contains considerable rugged terrain, most of it covered by virgin rain forest. It is an increasingly popular nature tourism destination for domestic and foreign tourists.

The authors' econometric analysis focused on domestic visitors. Data were collected in 1988 from 755 visitors. Visitation rate-demand functions were estimated and used to calculate economic surplus measures for tourists coming from each of 81 cantons (geopolitical regions) in the country. From their analysis, they estimated that the site was worth about $35 per visit for the 3,000 Costa Ricans who annually visit. Making a conservative assumption that site visits would be worth at least $35 to the foreign tourists as well, the authors calculated that the present value of recreation for both tourists groups was $2.5 million to $10 million, or about $1,250 per hectare. They concluded that since the acquisition price for additional reserve land was $30–100 per hectare, expansion of the reserve is economically justified based on recreational use alone.

For example, one could collect information from tourists on the number of birds, reptiles, and large mammals seen, and then use that information in conjunction with ratings of sites and number of visits to sites to determine the value of species diversity for tourists (Mercer and Kramer, 1992).

A more direct approach to estimating the value of biodiversity is the contingent-valuation method (CVM), which uses survey research methods to ask people directly what they are willing to pay for a change in the level of provision of a nonmarket public good. The results depend on an appropriate hypothetical market in which individuals have an opportunity to bid on the good in question. The advantage of CVM is that it can capture both use and nonuse values. Economists have used CVM to value individual endangered species such as humpback whales, whooping cranes, and bald eagles (Loomis and Walsh, 1986).

CVM research on valuing the protection of endangered species has shown that nonuse values generally exceed use values. People express a willingness to pay for programs to protect endangered species, apparently because they enjoy knowing the species will continue to exist in the wild. Another finding is that people are willing to pay more for familiar species, which suggests that the dollar estimates obtained in

CVM studies may be influenced by the information about biodiversity provided to people during surveying. People also bid more for species that are "cute and cuddly." Hence, a primate will be valued much more highly than a nematode.

Using CVM to place a value on individual species, however, is unlikely to be a fruitful way of measuring the full benefits of biodiversity protection. Instead, a more useful approach is likely to be the use of CVM to measure the perceived benefits of protecting intact ecosystems. But it is a big step to go from valuing a single, often well-known endangered species to the much more complex good known as biodiversity. Moreover, this approach may be fraught with uncertainty, because the elicited values will be sensitive to the information provided to survey respondents. Nevertheless, this work on valuing endangered species is a useful starting point for understanding the challenges of applying economic analysis to biodiversity. Given the growing level of policy directed toward conserving biodiversity, it is becoming increasingly important to expand research efforts on this and other benefit-cost methods to value biodiversity.

It might also be noted that because of the inherent difficulties associated with measuring the benefits of biodiversity protection, some have argued against using benefit-cost analyses at all, opting instead for the use of a "safe minimum standard" decision framework (Tisdale, 1990). This approach assumes that there are significant benefits associated with biodiversity protection but does not attempt to measure them. Indeed, it places emphasis on measuring the opportunity cost of maintaining minimally viable populations or ecosystems. The drawback of this approach is that by placing so much attention on only the cost side, it may bias decisions toward development and away from protection.

## CASE STUDIES OF NONMARKET VALUATION METHODS

The application of nonmarket valuation methods to measure the economic value of biodiversity is illustrated in the following two case studies. In the first study, the benefits and costs of establishing a new protected area in Madagascar were estimated for two different stakeholder groups: local residents and foreign tourists. In the second study, the value of biodiversity protection as a global environmental good was estimated for a group of U.S. residents who were queried about their WTP for a doubling of the total size of protected areas in tropical rain forests.

### Economic Impacts of a Park in Madagascar

Situated off the east coast of Africa, Madagascar is a 1,000-mile-long island with a population of more than 11 million. It has been singled out by the international environmental community as one of the ecologically richest countries in the world and one whose biological diversity is at great risk. All of the island's mammals, 80 percent of its plants, and half of its birds occur no where else in the world (World Bank, 1989). At the same time, Madagascar is one of the world's economically poorest countries, with a per capita income in 1990 of $230.

With international assistance, the Madagascar government is establishing a system of parks and reserves throughout the country. Given that this new management system is replacing a combination of open access and communally controlled access, will there be positive net benefits from these investments in protected areas? To begin to address this question, we and several colleagues have conducted research to quantify the benefits and costs of a new park being established in the eastern rain forest, called the Mantadia National Park (Kramer et al., 1993; Shyamsundar, 1993). To assess the economic impact of the park, it is necessary to determine how the park will affect the total value of the forest, which includes both use and nonuse values. The use values include what households derive from the flow of fuelwood, building poles, crayfish, fruits, medicine, and other products from the forest, as well as soil nutrients obtained from the forest by means of swidden (formerly called slash-and-burn) agriculture. Use values also include the value to foreign and domestic tourists who engage in recreation activities in the forest, to locals who use the forest for religious and cultural purposes, and to scientists who use the forest for biodiversity investigations. The nonuse values are primarily existence values: due to the rich habitat provided by the forests in Madagascar and the unusually high number of endemic species, many people may value forest protection in the country even though they never plan to visit the forests.

It was not possible to measure all of these values within the context of the study, so we focused attention on the economic impacts on local villagers and foreign tourists. The study had two major components: the use of CVM and opportunity-cost analysis to measure the benefits and costs of the park to local people, and the use of CVM and a recreation-demand model to measure the value to foreign tourists of the new national park.

The opportunity cost to villagers of losing access to the land set aside for the Mantadia National Park was determined by analyzing data from a socioeconomic survey administered to 351 households living in the vicinity. Cash-flow models were constructed for each village group in order to estimate income from agricultural and forestry activities. According to the prior dependence of each group on park land for forest products and shifting agriculture (based on analysis of aerial photographs), we estimated the income reductions resulting from lost access to those resources. The mean annual extent of these losses was estimated to be $91 per household. By adding up these costs for all households living in the vicinity of the park, and assuming a 3 percent rate of population growth and a 10 percent discount rate, we estimated the net present value of the opportunity costs over a period of 20 years to be $566,000 (see Table 8-1).

In addition to the opportunity-cost analysis, we also asked villagers directly about the losses imposed on them by the park. They were asked what level of compensation (in baskets of rice) would make them just as well off as they had been before the park was established. The results of the CVA indicate an annual average compensation of rice equal in value to $108 would make households as well off with the park as without. To cover all the people in the park area, this implies a one-time compensation of approximately $673,000. Of course, in practical terms it is unlikely that compensation would be in the form of rice or cash, but instead might be in the form of investments in local infrastructure, schools, or agricultural intensification projects.

Where would the funds for such compensation come from? One obvious source would be a user fee or tax on tourists coming to enjoy the new park (Dixon and Sherman, 1990). Working with data from a survey of tourists visiting a nearby small reserve, supplemented by data from a survey of tour operators, a recreation-demand model was used to estimate the potential benefits to foreign tourists, who in fact make up most of the visitors to parks in Madagascar (Mercer et al., 1993) The econometric analysis examined the allocation of trip choices to Madagascar and other international nature-tourism destinations as a function of travel costs, socioeconomic characteristics, and quality variables. (As in the study in Costa Rica described in Box 8–1, the estimated recreational value was for all attributes of the park, not just biodiversity.) The analysis indicated that the new park would be worth an average of $24 per tourist trip. Based on the conservative assumption that the same number of foreign tourists (3,900) would visit the new park as visit the

**Table 8-1** Summary of the economic analysis of Mantadia National Park in Madagascar

| Method used | Annual mean value[a] | Aggregate net present value |
|---|---|---|
| Estimates of welfare losses to local villagers from establishment of the park | | |
| Opportunity cost | $91 | $566,000 |
| Contingent valuation | $108 | $673,000 |
| Estimates of welfare gains to foreign tourists from establishment of the park | | |
| Travel cost | $24 | $936,000 |
| Contingent valuation | $65 | $2,530,000 |

[a]Annual mean value per household for estimates of welfare losses to local villagers; per trip for estimates of welfare gains to foreign tourists.

nearby forest reserve, the aggregate benefit to tourists during a 20-year period would be $936,000 (see Table 8-1).

The study also used CVM to query foreign tourists directly about the economic value of the park. Visitors were given information about the new park and then asked how much more they would have been willing to pay for their Madagascar trip if the park was a part of their travel itinerary. The average increase in WTP for their trip was $65. Assuming that current visitation patterns continue, the total additional WTP to visit the new park would be $2.53 million over a period of 20 years. The higher valuation estimate for the CVM approach may be due to its ability to capture nonuse as well as use values.

This case study shows that in-depth economic analysis of protected area projects can have implications for their design and implementation. Such information can be used, for example, to determine compensation for local people displaced from forest areas designated as national parks (see chapter 9). Also, economic valuation efforts can establish the value of a park as a nature tourism experience for foreigners, thus influencing external assistance for conservation programs and suggesting ranges of feasible user fees to generate funds for maintaining conservation activities in parks and buffer zones.

## Benefits to U.S. Residents from Tropical Rain Forest Protection

Public opinion polls and voluntary contributions to conservation organizations bespeak evidence for substantial public support for biodiversity protection. Another way of gauging this support is to use CVM. It has already been noted that research has demonstrated substantial existence values for various species and ecosystems, especially among residents of the developed world (Loomis and Walsh, 1986; Imber et al., 1991). Here, we summarize a national contingent-valuation survey used to assess the value that U.S. residents place on protecting rain forests (Kramer et al., 1996).

A mail survey was sent to a random sample of 1,200 U.S. residents between April and June 1992. After several follow-up mailings, the final response rate was 56 percent. The respondents were asked a number of questions about their attitudes toward environmental and other social issues, their familiarity with tropical forests, their socioeconomic characteristics, and their willingness to contribute to efforts to protect rain forests.

The respondents were familiar with tropical deforestation: 91 percent responded affirmatively to the question "Before today, have you ever read, heard, or seen TV shows about tropical rain forests?" and 81 percent claimed to be familiar with reasons for deforestation (see Table 8-2). Two thirds of the respondents answered "yes" to the question "Should industrialized countries help developing countries pay for pre-

*Table 8.2* Reponses to 1992 U.S. survey about knowledge of, visits to, and obligations to pay for rain forests

| | Yes | No |
|---|---|---|
| Any knowledge of rain forests | 91% | 9% |
| Knowledge of causes of deforestation | 81% | 19% |
| Previously visited a rain forest | 11% | 89% |
| Plan to visit a rain forest | 8% | 61% |
| Industrialized countries should help developing countries pay for preserving their rain forests | 67% | 33% |

*Source:* Kramer et al., 1996.

serving their rain forests?" This answer has important ramifications for the ongoing political debate about the role of industrialized countries in bearing some of the costs of environmental protection in developing countries. A follow-up question asked what percentage of the costs should be borne by the industrialized world, and the median response was 41 percent.

After providing arguments for and against tropical rain forest protection, we asked respondents if they would be willing to contribute to a United Nations–managed fund to protect an additional 5 percent of the world's tropical forests. On average, the respondents were willing to make a one-time payment of approximately $24 per household (see Table 8-3). Aggregating the households, this implies a WTP of nearly $2.2 billion. If one assumes more conservatively that only households with incomes of more than $35,000 could actually be convinced to contribute voluntarily or through taxes, the fund would be in the order

*Table 8-3* Estimated willingness to pay (WTP) by U.S. residents for protecting tropical rain forests

| Measure | Estimated Value |
|---|---|
| Mean WTP ($/household) | $24 |
| Total WTP | |
| All households[a] | $2,184,000,000 |
| Households with income > $35,000[b] | $780,000,000 |
| Households with income > $25,000[b] | $1,007,000,000 |

*Source:* Kramer et al., 1996

[a]Assuming 91 million households in the United States in 1989 (U.S. Bureau of Census, 1995)
[b]Income distribution in 1989 (U.S. Bureau of Census, 1993)

of $800 million. This could create a substantial global fund indeed, if households in other industrialized countries are willing to make similar-sized donations. For both methodological and policy-information purposes, it would be of interest to replicate this study in other countries with similar income levels to determine if the WTP for global environmental goods varies across countries for cultural or other reasons.

## NATIONAL ENVIRONMENTAL ACTION PLANS IN BIODIVERSITY PROTECTION

In recent years, an important instrument for promoting conservation of biodiversity in developing countries has been the development of National Environmental Action Plans (NEAPs). A NEAP, which is developed internally by individual countries, provides a long-term conceptual framework that encompasses various government sectors and links environmental and natural resource management goals into the country's national development program (World Bank, 1993b).

As an ongoing process, the NEAP defines a carefully scheduled action plan that focuses on policy changes, capacity building, institutional and legal reforms, development of human capital, environmental education, development of an information system, and investment projects to achieve better use of natural resources and management of the environment. In addition, the NEAP includes projects to correct existing environmental damages, such as industrial pollution, land degradation, and loss of water quality. The actions are designed to create incentives for conservation of natural resources and to build in-country capacity for environmental management. The NEAP is based on a participatory approach, involving local people, traditional institutions, nongovernmental organizations (NGOs), the private sector, the government, and donor agencies. The NEAP also provides an important basis for engaging the international community (e.g., through the Global Environmental Facility [GEF] and bilateral and multilateral assistance programs) in providing funding, training, and technical assistance for dealing with environmental degradation and pollution problems and for implementing projects to promote sustainable use and conservation of natural resources.

Some 25 developing countries have now prepared NEAPs, and many others are initiating the exercise. Box 8-2, drawn from Sri Lanka, illustrates the types of biodiversity activities included in NEAPs. A number of countries—including Madagascar, Mauritius, Seychelles, Sri Lanka, Egypt, and Ghana—have begun implementing their action plans (World Bank, 1993a). But more work is needed to strengthen the NEAP process, the content of the action plan, the economic analysis of proposed policies and investment programs, the implementation of the action plan, and the participation of local people, the private sector, and NGOs.

Madagascar's NEAP has strongly emphasized biodiversity conservation. Considered one of the megadiversity countries because of its high rates of species endemism, Madagascar has received large amounts

***Box 8-2*** Biodiversity conservation measures in Sri Lanka's national environmental action plan

- Assessment of existing protected areas to ensure that they represent the diversity of biological resources and ecosystems in the country and to identify other sensitive areas to be included in expanded protected area networks.
- Formation of a wildlife trust fund.
- Promotion of ecotourism by establishing incentives for involvement of the private sector.
- Preparation of a national wetland inventory and a master plan.
- Execution of conservation-management plans for six protected areas and other sites involving public and private sectors.
- Development of buffer zones, done with the involvement of local people.
- Implementation of a management plan to protect elephants.

of funding for biodiversity protection from the international community. Through its NEAP framework, Madagascar has established institutions and programs for conservation of natural resources and environmental protection. In 1991, the country established a Multi-Donor Secretariat to facilitate donor support to help in implementing the environmental action plan, which includes protection of 50 parks and natural reserves in several forest areas. More effort is needed to accelerate the pace of implementation, and priorities must be clearly defined (Falloux and Talbot, 1993; Larson, 1993).

The NEAPs can become important vehicles for promoting and financing conservation of biodiversity and protection of wildlife resources on a significant scale and for combining national efforts with international support through concessionary financing, grants, technical assistance, and research. However, securing adequate national and donor support to carry out the plans remains a major challenge. To the extent that the NEAPs include provisions for protecting biodiversity assets that have global significance, it may be possible to use funding from the GEF (see below).

## INTERNATIONAL FUNDING FOR BIODIVERSITY PROTECTION

### Debt-for-Nature Swaps

Debt-for-nature swaps are arrangements in which a conservation organization purchases a developing country's debt at a discounted value in a secondary debt market (Hansen, 1988); the purchased debt is then canceled in exchange for the debtor country's pursuit of an environmentally related activity. Often, the agreement is for the country to take local currency in the amount of the debt's face value and invest it in conservation activities managed by local organizations. Since a number of developing countries have both large debts and significant amounts of tropical forests, organizations such as the World Wildlife Fund and Conservation International have seized this opportunity to facilitate habitat preservation at a relatively low cost. The arrangements have sometimes been criticized, however, as foreign impingement on debtor countries' sovereignty.

Conservation International arranged the first debt-for-nature swap, in Bolivia in 1987 (McNeely, 1988). Since then, swaps have been arranged in ten other countries: Madagascar, Zambia, Costa Rica, Dominican Republic, Ecuador, Guatemala, Jamaica, Mexico, Philippines, and Poland. As of 1992, $99 million in debt had been purchased by a variety of conservation and other organizations for a cost of $17 million. In turn, governments had set aside $62 million in conservation funds. Some, but not all, of these funds were devoted to biodiversity protection. Examples include funds to purchase park land or manage parks in Costa Rica, Philippines, and Jamaica and to establish conservation data systems in Ecuador and Mexico (WRI et al., 1992).

The transfer of funds from developed countries to developing countries through debt-for-nature swaps represents an innovative means to capture WTP for forest protection. Brown et al. (1993) examined 23 swaps and used the available information to calculate an implicit price per hectare of protected land. They calculated a range of 18 cents to $11 per hectare and suggested a benchmark value of about $5 per hectare. They concluded that these implicit prices "are capturing only part of the rich world's existence values for these assets" (p. 21).

## Global Environment Facility

The Global Environment Facility (GEF) is an effort to provide international cooperation and financing for large-scale environmental projects in developing countries. The facility is implemented jointly by the United Nations Development Programme (UNDP), the United Nations Environment Programme (UNEP), and the World Bank. Established in 1990 with initial funding of $1.5 billion for three years, the GEF has four major program areas: reducing global warming, preserving biological diversity, protecting international waters, and slowing depletion of the ozone layer. Approximately 20 countries have contributed to the fund. The World Bank administers the fund and has responsibility for appraising and supervising projects. Among the projects already approved are biodiversity projects in East Africa and the South Pacific and a project to protect tropical forests in the Congo.

The GEF is emerging as the primary international financing arrangement for biodiversity conservation efforts. It is also serving as at least an interim funding instrument for implementing the goals of the Convention on Biological Diversity, agreed upon at the Earth Summit, held in Rio de Janeiro in June 1992. While debt-for-nature swaps have represented a modest attempt to transfer rich countries' financial resources to the developing world for forest protection, the GEF is a much more ambitious effort to capture the value of global benefits from biodiversity protection and distribute the financial resources in the poorer countries. Furthermore, it includes a planning process that attempts to assess the relative merits of biodiversity protection efforts in different parts of the developing world and allocate financial resources accordingly. For biodiversity projects, the GEF serves to raise the economic return to protected area projects. To compensate for foregone development alternatives, the GEF transfers global resources to developing countries to pay for the costs of protection (Brown et al., 1993). This transfer enables the increased protection of biodiversity, which in turn generates increased global benefits.

## TRUST FUNDS FOR SUSTAINABLE FINANCING OF PROTECTED AREAS

National Environmental Action Plans and international financing agreements can raise awareness about biodiversity conservation and provide initial funding for projects. However, protected areas need a permanent source of funding if they are to play their role as long-lasting storehouses of biodiversity. Many conservation projects are funded as a one-time grant or a several-year project (e.g., 5–10 years) by a bilateral or multilateral agency, with the expectation that the host country will later pay the project's operating and maintenance costs. Given shifting priorities and the relatively meager financial returns generated by parks, it is not surprising that protected areas are often underfunded by national governments. Even when there is a continuing strong commitment to conservation, economic shocks such as recessions, wars, and fluctuations in oil prices can lead to a downturn in funds available to operate protected areas. The effects of fluctuating and declining funding for biodiversity protection are seen in the disturbing statistics on the state of protected areas reported in Chapter 4 of this book.

A trust fund is money set aside under the legal control of trustees to benefit a group of beneficiaries. The trustees have a fiduciary responsibility to the trust's beneficiaries and can be sued by the beneficiaries if the terms of the trust are not honored. Trusts are a common-law concept, but they exist under other legal systems as the equivalent of a charitable foundation. (For an example of legislation to create a conservation trust fund in Madagascar, see Cantin, 1994.)

From the perspective of the economics of protected areas, the most attractive feature of trust funds is that they create an income stream that lasts far longer than the financing mechanisms of traditional development and conservation projects. Another advantage is that their governance (usually a board of trustees) allows an opportunity for representation by diverse stakeholders, including conservationists, indigenous people, and government agencies. Also, trust funds are more flexible than traditional financing mechanisms, so they are better able to respond to experimental projects and to expand the absorptive capacity of implementing agencies (Wells, 1991).

The best-known conservation trust funds that have benefited protected areas have been established through debt-for-nature swaps (e.g., the Foundation for the Philippine Environment and Bolivia's National Fund for the Environment). However, trust funds can be financed in a variety of ways, including private contributions, donations from NGOs, and funding by host country governments. Belize's Protected Areas Conservation Trust is funded by a fee of $20 per person levied on foreign tourists and by site-entry fees and the sale of recreation-related licenses. The GEF is supporting several conservation trust funds and may considerably expand the use of this financing tool in the future.

## RECOMMENDATIONS

### Tradeoffs in Protecting Biodiversity

The drive by developing countries for economic development has undoubtedly placed them at an important crossroads. Each country can accelerate economic growth under prevailing conditions and modes of production through increased investments, advanced technology, and more intensive use of natural resources. This option will allow a country to achieve rapid economic advancement, especially in the short term and medium term. But given current trends in environmental degradation, pollution, and depletion of natural resources, the country will most likely continue to experience serious environmental consequences—including loss of tropical forests—with significant social costs and burden on future generations. On the other hand, a country can promote sustainable development by integrating conservation values and approaches into its development programs. This approach will enable the country to improve human well-being while protecting its environment and natural resources. Ultimately, the state of a developing country's environment will determine the fate of its long-term economic and social progress.

Each developing country's development challenge will include hard choices and tradeoffs between immediate economic gains from consumption and long-term benefits from conservation of natural resources and protection of the environment. The choices made will have economic relevance as well as ecological and ethical implications. As noted earlier, biodiversity and environmental services provided by forests are public goods not valued and traded in the market; therefore, decisions relating to the use of these resources should not be based strictly on market criteria. To reconcile private and societal interests, countries must weigh ecological, ethical, and political considerations as well. For instance, how much area of tropical rain forests should a country protect for conservation of biodiversity? What is an acceptable level of risk of loss of species, ecosystems, and ecosystem functions? To what extent must the current generation sacrifice to maintain biodiversity-related options for future generations?

Individuals, special-interest groups, and political institutions respond differently to environmental problems and imperatives. Within the context of ongoing economic and political changes in developing countries, these elements will increasingly affect choices between conflicting objectives. The market will facilitate some choices, but government intervention will be necessary in some instances to deal with market failures in order to protect the long-term interests of society. For example, the allocation of land among competing economic activities (such as agriculture, mining, ranching, timber production, and wildlife management) will require both the market and the public sector to ensure that economic growth and environmental protection aims are achieved. But land allocation will also involve tradeoffs. How much land should be converted from existing forests and woodlands to agriculture? Which areas should be targets, and will these areas support sustainable agriculture? Which forest areas should be protected? Should each tropical country protect the same proportion of its remaining forests? How can protected areas be justified to poor residents? How can protected areas be financed? Should former forest exploiters be compensated? Another set of issues involves transboundary questions. Should a developing country invest in protecting natural forests and their biodiversity when most of the benefits derived go to people in other countries? How should the developing countries be compensated? How can it be assured that they pay their fair share and not place all of the burden on the rest of the world?

The capacity of developing countries to deal with choices and tradeoffs will improve immensely with better databases, improved scientific knowledge, and analysis of environmental problems and consequences. Better information is also needed on the broad implications of environmental policies in terms of efficiency, equity, and sustainable development. With improved knowledge, countries would have a better understanding of the temporal and spatial dimensions of biodiversity losses and environmental degradation. However, lack of data and knowledge should not be a barrier to action that could be based on lessons from best practices and pilot programs, as well as on localized knowledge and experience. To this end, a lack of government capacity should be offset, in the short term and medium term, by "second-best" approaches supported by the international donor community.

## A General Strategy

Tropical forest conservation and development efforts should be directed toward stabilizing existing forests through sustainable management of the forest ecosystems and through reducing destructive deforestation. Under sustainable management, development priorities must focus on how forests can be integrated into the NEAP process and land-use plans for meeting the demand for timber, fuelwood, fodder, and other wood and nonwood products; for providing environmental services and for conservation of biological resources; for dealing with land degradation and stabilization of watersheds; for increasing agricultural productivity; for meeting subsistence energy needs; for meeting the needs of the indigenous people; and for addressing regional and global concerns. While the focus of this book is on protected areas, protection will be effective in the long run only if these and other objectives and needs are met through a coherent set of policies and markets operating in the larger landscape surrounding protected areas.

To this end, the following four-point strategy for environmentally sustainable development of tropical forests is proposed:

1. *Increase investments in forest development and conservation.* Although developing countries are already experiencing some increase in investments in forestry, development efforts and capital must increase significantly for these countries to achieve desirable forestry development and conservation goals. Natural regeneration and reforestation have not kept up with the rate of deforestation. A strong effort is now needed to stabilize forest areas through sustainable management of natural forests. In addition, there is a need to expand reforestation and afforestation (planting trees in new areas) for production and protective purposes, as well as to plant trees on integrated production systems (farm and range lands) for sustainable land use. Successful protection of areas set aside for biodiversity conservation will occur only if there are other productive areas for meeting the material needs of the human population.
2. *Increase ecological, economic, and policy research.* The threat of depletion of tropical forests in general and protected areas in particular calls for improved knowledge, more efficient resource planning and management, incremental response and adjustments, and better understanding of the cause-and-effect relationships of these problems in a national and global context. Solutions must meet the needs and priorities of the developing countries, while recognizing the importance of efficiency, equity, intergenerational considerations, and global concerns. Solutions must also address cost and compensation aspects to provide incentives for remedial or preventive actions.

3. *Establish policies and priorities for ensuring improved management and conservation of forest resources.* Development priorities reflecting the productive and protective functions of forests should be established to address how some of the existing forests can be managed sustainably for single or multiple uses, how the present rate of destructive forest removal can be mitigated, how forests can increase agricultural productivity and stabilize degraded lands and watersheds, and how reforestation and afforestation can be expanded significantly to provide environmental protection and to meet growing demands for timber, fuelwood, and fodder. To this end, adjustments and development efforts need to focus on forest incentive policies and nonforest policies, on management systems for technology and forests, on development of institutions and personnel, on protection of critical forest ecosystems for conservation of biological resources and natural habitats, on the roles of public and private sectors and of individuals in forest use and management, and on global initiatives to arrest deforestation.
4. *Increase international financial and technical assistance.* To support developing countries in achieving forestry conservation and development goals, the world community must provide substantial financing and technical assistance, especially for expanding protected areas for conservation of biodiversity. External assistance must support significant policy reforms, institution building, personnel development through training and research, and development of information systems.

## Action and Research Agenda

To implement this four-point strategy and address the biodiversity crisis, the following action and research recommendations are proposed.

At the local level:

- Initiate a series of studies with different ecological, social, and economic conditions to determine the benefits and costs of protected area projects to local people. Identify the stakeholders involved and estimate the gains and losses.
- Use the studies above to determine a fair level of compensation, or a substitute set of economic activities, for people living adjacent to protected area projects. Because it may not be possible to undertake detailed economic studies for each proposed project, compensation should be based on benchmark projects with similar economic land uses and numbers of affected people.
- Involve local residents in the design and management of protected areas and associated buffer zones to maximize their sense of involvement and stake in the projects. Select appropriate means of compen-

sation (e.g., health clinics, community forestry projects) based on the desires of local residents.
- Clarify or modify property rights to allow local residents to expand agroforestry, participate in extractive reserves, and have access to productive forests owned by the public sector.

At the national level:

- Reform forestry and agricultural policies that create a bias against maintaining healthy forested landscapes. Design and implement policies that will make agriculture more intensive rather than more extensive, especially in areas adjacent to protected forests.
- Establish trust funds at the country level (from debt-for-nature swaps, the GEF, tourist user fees, and endowments) to ensure a sustainable source of operating and maintenance funds for protected areas and buffer zones. These trust funds should be administered by an independent body consisting of representatives of the public and private sectors, local communities, and nongovernmental organizations.
- Increase the level of scientific and economic analysis on which NEAPs are based, to ensure that biodiversity conservation needs are properly assessed in selecting projects and setting their priorities.
- Strengthen the NEAP process, define an action plan for protecting biodiversity within the NEAP framework, and establish a mechanism for donor assistance and coordination.
- Improve database and information systems through resource inventories, land surveys, and Geographic Information Systems (GIS), and establish a monitoring system for biodiversity assets.
- Implement comprehensive national land-use policies and plans that establish environmentally and economically sound options for using different types of lands. Through such policies and zoning, forests can be classified for various uses (e.g., preservation forests, protection forests, production forests, extractive reserves, forests for conversion to agriculture, forest areas for mining) with a long-term focus.

At the international level:

- Establish an international target for protecting and managing a minimum of 10 percent of tropical rain forests by the year 2005, using national and international resources to fund the protected areas and associated buffer zones.
- Establish an international target of creating 100 million hectares of new forest by the year 2005 through tree-planting programs and reforestation.
- Establish a world commission on natural forests, including tropical forests, to adopt a common agenda for sustainable management of these resources, develop an action plan (including policy and institutional reforms and a priority investment program), and promote consensus among nations of the world community.

- Expand the rich countries' contributions to the GEF and emphasize its role in financing the protection of globally important biodiversity.
- Support applied research (on ecosystem functions, genetic material, management systems, silvicultural practices, valuation of forest goods and services) and training in tropical forest countries.
- Ensure that all donor-financed projects that deal with tropical forests conduct comprehensive environmental, biological, and social assessments.
- Intensify donor efforts by supporting local programs in the following 15 tropical countries, which account for more than 75 percent of tropical rain forests: Bolivia, Brazil, Cameroon, Colombia, Congo, Ecuador, Gabon, India, Indonesia, Madagascar, Mexico, Myanmar, Papua New Guinea, Venezuela, and Zaïre.

## LOOKING AHEAD

The urgency of the tropical biodiversity crisis requires a major infusion of funds and a sharper policy focus if even a modest amount of the remaining resources are to be protected. The public's perception of this concern has grown, and this presents a political opportunity to push for sustainable forest management and increased forest protection in the tropics. Countries are faced with the difficult task of stabilizing forests by arresting deforestation, managing natural forests on a sustainable basis, increasing forest resources through reforestation and afforestation, and improving park and reserve systems. Developing countries have taken steps to address this task, but the level of effort needs to increase substantially.

The world's population is expected to double over the next 30 years, with most of this increase occurring in the developing countries. More than 1 billion people presently live under poverty conditions. These trends will continue to place pressure on existing tropical forests because of increasing demand for forest products and lands for agriculture. Developing countries are hard-pressed to augment food production and to improve the economic and social conditions of the poor through development efforts. Extensive agriculture will to a large extent rely on conversion of forest areas into arable lands.

It is against this challenge that the developing countries and the industrialized nations should assess priorities in dealing with biodiversity protection and in establishing and adequately funding new protected areas. As an important element of its policy, each country should set aside and maintain a significant portion of its total forest lands as protected areas for conservation of biodiversity for the benefit of present and future generations. If the world community wants tropical countries to protect more of their natural forests for conservation of biodiversity, it should recognize that those countries will incur opportunity costs from foregone development activities. These costs should be borne by both those nations where the protection activities occur and the citizens of the global community that benefit from protection. Currently, most of those benefits appear to lie outside of the tropical countries, although this situation will change as incomes rise and the demand for environmental quality increases in the developing world. The increased international funding for biodiversity protection through the GEF is a step in the right direction for facilitating international cost sharing, but the amount of funding available pales in comparison to the tasks remaining.

## NOTES

We benefited from discussions with Joe Aldy, Paul Ferraro, Evan Mercer, Nick Salafsky, Kathryn Saterson, Madhu Rao, Priya Shyamsundar, Carel van Schaik, and Jan Wind.

## REFERENCES

Binswanger, Hans. 1989. *Brazilian policies that encourage deforestation in the Amazon*. Environment Department Working Paper No. 96. Washington, D.C.: World Bank.

Bockstael, Nancy E., Kenneth E. McConnell, and Ivar Strand. 1991. Recreation. In *Measuring the demand for environmental quality*, Jon B. Braden and Charles D. Kolstad (eds.), Chap. 8. New York: North-Holland.

Botkin, D., and Lee M. Talbot. 1992. Biological diversity and forests. In *Managing the world's forests*, Narendra Sharma (ed.), 47–74. Dubuque, Iowa: Kendall/Hunt Publishing.

Brown, G., and R. Mendelsohn. 1984. The hedonic travel cost method. *The Review of Economics and Statistics* 66:427–33.

Brown, Katrina, David Pearce, Charles Perrings, and Timothy Swanson. 1993. *Economics and the conservation of global biological diversity*. Working Paper No. 2. Washington, D.C.: Global Environment Facility.

Cantin, Egide. 1994. Draft legislation for the creation of a National Environmental Endowment Fund in Madagascar. Arlington, Va.: Winrock International Environmental Alliance.

Dixon, John A., and Paul B. Sherman. 1990. *Economics of protected areas*. Washington, D.C.: Island Press.

Falloux, Francois, and Lee M. Talbot. 1993. *Crisis and opportunity: environment and development in Africa*. London: Earthscan Publications.

Freeman, A. Myrick. 1993. Nonuse values in natural resource damage assessment. In *Valuing natural assets: the economics of natural resource damage assessment*, Raymond Kopp and V. Kerry Smith (eds.), 264–303. Washington, D.C.: Resources for the Future.

Hansen, Stein. 1988. *Debt for nature swaps: overview and discussion of key issues*. Environment Department Working Paper No. 1. Washington, D.C.: World Bank.

Imber, D., G. Stevenson, and L. Wilks. 1991. *A contingent valuation study of the Kakadu Conservation Zone*. Research Paper No. 3. Canberra, Australia: Resource Assessment Commission.

Kramer, Randall A., Robert Healy, and Robert Mendelsohn. 1992. Forest valuation. In *Managing the world's forests*, Narendra Sharma (ed.), 237–67. Dubuque, Iowa: Kendall/Hunt Publishing.

Kramer, Randall A., Evan Mercer, and Narendra Sharma. 1996. Valuing tropical rain forest protection using the contingent valuation method. In *Forestry, economics and the environment*, W. L. Adamowicz, P. C. Boxall, M. K. Luckert, W. E. Phillips, and W. White (eds.), 181–194. Wallingford, U.K.: Centre for Agriculture and Biosciences International.

Kramer, Randall A., Narendra Sharma, Evan Mercer, and Priya Shyamsundar. 1993. *Environmental valuation of tropical forest protection in Madagascar.* Final Project Report to the World Bank. Durham, N.C.: Duke University, Center for Resource and Environmental Policy Research.

Larson, Bruce A. 1993. Changing the economics of environmental degradation in Madagascar: lessons from the National Environmental Plan process. Paper presented at the Northeast Universities Development Consortium Conference, Williamstown, Mass., 15–16 October.

Lindberg, K. 1991. *Policies for maximizing nature tourism's ecological and economic benefits.* Washington, D.C.: World Resources Institute.

Loomis, J. B., and R. G. Walsh. 1986. Assessing wildlife and environmental values in cost-benefit analysis: state of the art. *Journal of Environmental Management* 22:125–31.

McNeely, J. A. 1988. *Economics and biological diversity: using economic incentives to conserve biological resources.* Gland, Switzerland: IUCN (World Conservation Union).

Mercer, D. E., and R. A. Kramer. 1992. An international nature tourism travel cost model: estimating the recreational use value of a proposed national park in Madagascar. Paper presented at the annual meeting of the Allied Social Science Associations, New Orleans, La., 5 January.

Mercer, D. E., Randall A. Kramer, and Narendra Sharma. 1993. Estimating the demand for rain forest tourism. Unpublished manuscript, Duke University, Center for Resource and Environmental Policy Research.

Principe, Peter P. 1991. Valuing the biodiversity of medicinal plants. In *Conservation of medicinal plants*, O. Akerele, V. Heywood, and H. Synge (eds.), 79–124. Cambridge: Cambridge University Press.

Randall, Alan. 1986. Human preferences, economics, and the preservation of species. In *The preservation of species: the value of biological diversity*, Bryan G. Norton (ed.), 79–109. Princeton, N.J.: Princeton University Press.

Rowe, R., N. Sharma, and J. Browder. 1992. Deforestation: problems, causes, concerns. In *Managing the world's forests*, N. Sharma (ed.), 33–45. Dubuque, Iowa: Kendall/Hunt Publishing.

Sedjo, Roger. 1992. Preserving biodiversity as a resource. *Resources* (Winter): 26–29.

Shyamsundar, Priya. 1993. Economic implications of tropical rain forest protection for local residents: the case of the Mantadia National Park in Madagascar. Ph.D. diss., Duke University, School of the Environment.

Smith, V. K. 1990. Can we measure the economic values of environmental amenities? *Southern Economic Journal* 56 (4): 19–32.

Tisdale, Clem. 1990. Economics and the debate about preservation of species, crop varieties and genetic diversity. *Ecological Economics* 2:77–90.

Tobias, D., and R. Mendelsohn. 1991. Valuing ecotourism in a tropical rainforest reserve. *Ambio* 20 (2): 91–93.

U.S. Bureau of the Census. 1993. *Income and Poverty*. CD-Rom database. Washington, D.C.: U. S. Bureau of the Census.

U.S. Bureau of the Census. 1995. *Current Population Reports*. Series P-20. Washington, D.C.: U. S. Bureau of the Census.

van Schaik, C., N. Salafsky, R. Kramer, and P. Shyamsundar. 1992. *Biodiversity and tropical forests: monetizing and managing an elusive resource*. Durham, N.C.: Duke University, Center for Tropical Conservation.

Vincent, Jeffrey R., and Clark S. Binkley. 1992. Forest-based industrialization: a dynamic perspective. In *Managing the world's forests*, N. Sharma (ed.), 93–137. Dubuque, Iowa: Kendall/Hunt Publishing.

Wells, Michael. 1991. *Trust funds and endowments as a biodiversity conservation tool*. Environment Department Divisional Working Paper No. 1991-26. Washington, D.C.: World Bank.

World Bank. 1989. *Madagascar: adjustment in the industrial sector and an agenda for future reforms*. Report No. 7784–MAG. Washington, D.C.: World Bank.

World Bank. 1993a. *Ecologically sensitive sites in Africa*, Vol. 2: *Eastern Africa*. Washington, D.C.: World Bank.

World Bank. 1993b. *The World Bank and the environment*. Washington, D.C.: World Bank.

WRI, IUCN, and UNEP [World Resources Institute, World Conservation Union, and United Nations Environment Programme]. 1992. *Global biodiversity strategy: guidelines for action to save, study, and use earth's biotic wealth sustainably and equitably*. Washington, D.C.: WRI, IUCN, and UNEP.

# 9

# Compensation and Economic Incentives: Reducing Pressure on Protected Areas

*Paul J. Ferraro and Randall A. Kramer*

Although the global social benefits of establishing protected areas in tropical rain forests may outweigh the total costs, the local private costs of restricting access to an important resource may be relatively substantial for residents and communities. The imbalance between costs accruing at the local level and benefits accruing at the national and international levels has raised questions about whether people living in or near protected areas ought to be compensated for their losses, and if so, how compensation should be made.

The issue of compensating residents for lost resources has been discussed, implicitly or explicitly, in many treatments of the relationship between protected areas and local people, as well as in treatments of externalities. (Economists define externalities as actions of consumers or producers that affect the well-being of others in a way that is not reflected through prices or economic transactions.) In the literature on compensation, there is a large difference of opinion on whether compensation should be paid to victims of negative externalities, which include such things as the pollution of air or water and the siting of hazardous waste dumps.

A number of studies have argued for compensation of those people subject to negative externalities, at least in particular situations or through particular mechanisms (Johnson, 1977; O'Hare, 1977; Western, 1982; Knetsch, 1983; Ward, 1986; Tietenberg, 1988; Hodge, 1989; Sullivan, 1990, 1992; Barnett, 1991; Burrows, 1991; McNeely, 1991; Miceli, 1991; Farber, 1992; Pollot, 1993). Other authors, mainly economists, have argued equally persuasively against compensation in many or all situations (Knetsch, 1983; Blume et al., 1984; Baumol and Oates, 1988). Most of the differences of opinion derive from differences in the context of the case examined, the assumptions made, the criteria used for judging the desirability of outcomes, interpretations of relevant laws, and the proposed mechanism for compensation. In the context of protected areas, most authors have argued in favor of compensating residents (e.g., Western, 1982; Barnett, 1991; McNeely, 1991).

A unique best choice regarding compensation is not indicated in economic and political theory. Few protected area projects have attempted large-scale compensation initiatives; thus, there are few field examples to guide the discussion. In this chapter, we outline the various arguments for and against compensation as they specifically relate to the establishment of protected areas. The word "compensation" here indicates an explicit payment to residents equal to the value of their opportunity costs (i.e., the costs of foregone alternatives) resulting from restricted access to the protected area's resources. Compensation can take the form of cash payments, in-kind substitutes, infrastructure development, provision of social services, or the introduction of alternative production technologies. The focus of this chapter is on people living within or adjacent to protected areas whose daily subsistence and commercial needs have traditionally been met by using the area's resources, rather than on the larger-scale commercial users of these resources (e.g., timber companies, mining companies). The chapter ends by describing the ideal components of an approach that can ensure that residents are not worse off after the establishment of a protected area while at the same time generating local support for conservation.

## CRITERIA FOR JUDGING THE NECESSITY OF COMPENSATION

### Legality

In general, to compensate individuals for a lost benefit stream is to recognize that the individuals have some right to that benefit stream. Thus, when evaluating the necessity of compensation, one should first consider whether the law dictates that compensation is required. In many developed countries (e.g., the United States, England, Canada, New Zealand, and Norway), there are conditions under which government expropriation of resources or resource attributes is deemed a "taking" and the government must pay "just compensation" to the owners or users (Todd, 1976; Ward, 1986; Willis et al., 1988; Korsmo, 1991; Pollot, 1993). Some developing countries (e.g., Malaysia) also have similar laws (Knetsch, 1983). Some nations have specific laws governing the expropriation of tribal or indigenous lands (e.g., the United States, Australia, Bolivia), for which compensation is usually required (Holt, 1988; Yapp, 1989; Nash, 1993). In most developing countries, more than 80 percent of protected forest area is owned by the government (Gillis, 1991). In countries with large endowments of tropical forest, almost 100 percent of all natural forests is government owned (e.g., Indonesia, Malaysia, Ghana) (Gillis, 1991). The legal codes of most countries do not require compensation when the status of a public resource is changed.

If residents adjacent to protected areas do not have title to the land on which they live, they may have had certain legal rights to the resources on that land, which would strengthen their claim for compensation. However, even if the law does not indicate that compensation should be paid or if the residents had few legal rights to the resource prior to the protected area's establishment, there may be other reasons for compensating residents for the costs they incur.

## Efficiency

Most economic analyses of the allocation of land to a protected area would use the criterion of "efficiency" to judge the desirability of the allocation. Economists define an allocation of resources as efficient if the net benefit from the use of those resources is maximized by that allocation. This definition of efficiency has its roots in the concept of potential Pareto-improvement (PPI). A PPI is a change in the allocation of resources that could make, *after compensation,* at least one person better off and no one worse off. (For example, the maximum allowable herd size on public land is reduced by 10 percent, and the increase in benefits to other users of the land is larger than the decrease in benefits to the livestock owners. The gainers gain more than the losers lose, and thus the other users of the area could compensate the livestock owners. Reducing the allowable herd size by 10 percent is therefore a PPI.) At the societal level, an efficient allocation does not require compensation because society as a whole is better off.

Figure 9-1 depicts the optimal allocation of land to a protected area in a region. The y-axis measures dollars and the x-axis measures hect-

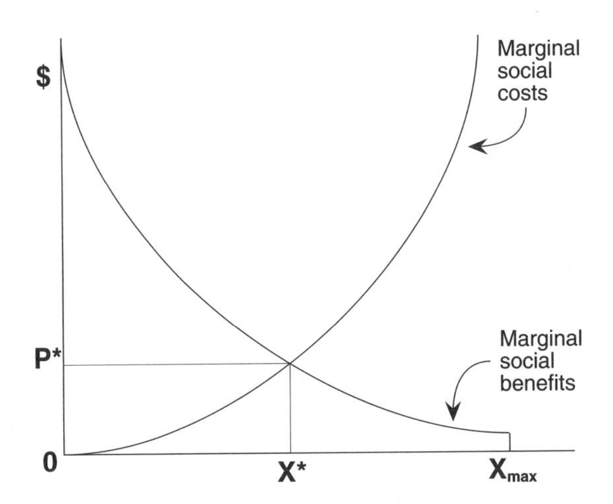


***Figure 9-1*** Costs and benefits of allocating land to protection.

ares of land allocated to protection, with $X_{max}$ being the maximum amount of land available in the region. If one assumes that only residents incur costs from protection, the social marginal cost curve can be interpreted as the marginal cost to local residents from supplying hectares of land for protection. If only nonresidents enjoy net benefits from protected areas, the social marginal benefit curve can be interpreted as the nonresident demand curve for hectares of protected land. $X^*$ is the economically efficient quantity of protected area where social marginal benefits equal social marginal costs.

If one assumes, for the moment, that the government is interested in maximizing social economic well-being and determines that $X^*$ is the amount actually allocated to the protected area, then compensation to residents is not required for efficiency. The efficient point has been chosen. The gainers could compensate the losers, but the efficiency criterion does not require it.

Since the government simply decreed the protected area, consumers of protected area benefits do not actually have to pay $P^*$ for each protected hectare. Thus, the entire area under the demand curve up to $X^*$ is consumer surplus (i.e., the difference between the amount consumers were willing to pay for the quantity demand and the amount they actually had to pay). Residents are left with only costs. Thus, although the allocation is efficient, it is highly inequitable—and this has been the cause of many conflicts worldwide between protected areas and local residents.

At first glance, it would seem that if there were some mechanism to force the beneficiaries to pay $P^*$ for each hectare protected, then one could maintain efficiency and compensate residents for their losses—and even give them some additional payment above and beyond their losses. But ecosystem protection is a public good subject to nonrival and nonexclusive consumption. (Nonrival consumption refers to consumption that cannot be divided among individuals; e.g., one citizen's consumption of national defense, does not diminish the amount available for others to consume. Nonexclusive consumption refers to the difficulty or impossibility of excluding nonpayers from using the good.) Because it is not possible to exclude nonpayers from enjoying some of the benefits of the protected area (e.g., Europeans who gain existence values for species protected in the tropics), there is the classic "free-riding" problem that causes markets to underprovide public goods. Hence, the price mechanism is not a workable way to finance protected areas, and governments must use tax mechanisms to force beneficiaries to pay. But the use of tax mechanisms to collect the "willingness to pay" (WTP) reflected in the marginal social benefits curve is problematic in this case, because many beneficiaries live in other countries. (See chapter 8 for a discussion of international transfer of funds to finance protected areas.)

Another way to redistribute some of the consumer surplus to residents would be to ask beneficiaries to pay their WTP for $X^*$ hectares voluntarily to a fund for local residents. This is effectively what happens now in many conservation projects in developing countries. When a nongovernmental organization sets up a protected area project, it is effectively representing its members and paying their WTP for the protection of habitat. A similar comparison can be made for protected area projects funded by multilateral and bilateral governmental donors; they are representing their taxpayers and paying their taxpayers' WTP for the protection of habitat.

Because of nonexclusivity in the consumption of ecosystem protection, free-riding behavior will decrease the effectiveness of this voluntary approach. Since the price bill exceeds the total damage bill (i.e., the area of $0P^*CX^*$ is greater than the area of $0CX^*$), enough people may pay so that the residents will be compensated for many of their losses. In general, however, it appears that the funds currently generated by donor organizations are insufficient given world conservation and compensation needs.

A number of authors have argued that requiring a government to pay the full cost of its projects will increase efficiency (Johnson, 1977; Ervin and Fitch, 1979; Posner, 1986). They maintain that if land is costless, the government may take land when costs outweigh benefits. The authors then claim that if the government must pay $P^*$ (or the area of $0CX^*$) to residents, there would be an incentive to protect only $X^*$.

However, the conclusion of these authors does not necessarily follow in all situations. In the case of protected areas, the conclusion ignores the direct costs of managing and protecting the area and the opportunity costs that certain powerful stakeholders (e.g., commercial timber interests) may bring to the government's attention. Moreover, there are three reasons why the government may choose to take land when costs outweigh benefits: the government is not acting as a maximizer of social economic well-being, the government does not have enough information to calculate $X^*$, or the government is not using efficiency as the sole criterion when making decisions about protecting biological resources.

If the first is the primary reason, then payment of $P^*$ will not necessarily make the government allocate $X^*$. A government acting as a maximizer of social well-being equates the marginal benefit of providing another unit of public good with the marginal cost of providing that unit. But if the government is not acting as a social well-being maximizer, then by definition it is not making such calculations. Thus, making it pay for the land is not likely to make it choose $X^*$.

Making the government pay may induce it to reduce the amount of land allocated to protected areas, but it may then allocate too few hectares to protection (i.e., less than $X^*$). This appears to have happened in Norway (Korsmo, 1991). In 1986, a new law mandated full compensation to owners of land subject to expropriation or regulation for conservation purposes. The law resulted in a decrease in the establishment of new protected areas at the same time that many scientific reports were arguing for an increase. In any case, making a nonoptimizing government pay for the land is unlikely to result in the efficient choice of $X^*$.

If the government is behaving like a social well-being (efficiency) maximizer, but lacks full information on costs and benefits, then it could not know the true value of $P^*$ to pay for each hectare, especially with the imperfect land markets in many developing countries. Making governments pay some price, however, may force them to collect more information on costs and benefits in an attempt to maximize the conservation value of their scarce funds. In making this calculation, the government must include the costs of obtaining the information.

In many cases, the government may have "overallocated" land to protection because efficiency was not the only criterion used to make the decision. Political, ecological, or ethical criteria may also have been part of the decision-making process. If this were the case, then the payment of $P^*$ by the government would not necessarily have resulted in a socially optimal allocation of land to protection.

In conclusion, it appears that it would be difficult to compensate residents for all their costs and still maintain efficiency. However, as mentioned in the discussion, the inequitable distribution of costs and benefits has been a focal point for many advocates of compensation. One could argue that it may be best to sacrifice some efficiency for equity.

## Equity and Ethics

Worldwide, many people are concerned about vertical equity. Vertical equity focuses on the way that the net benefits of a particular resource allocation are distributed among different income-level groups. Thus, it may be possible that a redistribution of wealth through compensation could be socially optimal. The resulting loss in efficiency could be made up for by the gain in social well-being that results from an increase in vertical equity. Given the potentially large amount of benefits from conservation, it is unlikely that the loss to the global society from inefficient compensation will be very significant. Moreover, any loss is likely to be widely distributed, so a small set of individuals will not have to bear the costs from the loss of efficiency.

The recent demands of developing countries for income transfers from the developed countries in order to pay for conservation (Speth, 1990; Haas et al., 1992) mirror similar demands that are arising at the local level. People living near protected areas in developing countries (or their advocates) claim that it is unfair for the residents of developed countries, who destroyed substantial amounts of their resource base to develop economically, to demand that local residents forego benefits in order to help conserve the scarce remaining biological resources in the world. The ecosystems designated for protection are valuable precisely because of the prior depletion of biodiversity elsewhere, regionally and globally. The residents around protected areas in developing countries are therefore being asked to forego benefits because people elsewhere have already depleted their local biodiversity. Thus, residents in the region of a protected area are being asked to incur costs that the majority of humans did not have to bear. One could argue that in order to mitigate the unfairness of this situation, compensation should be paid to the residents.

One could also argue for compensation on ethical grounds. Nonpayment of compensation implies that residents have absolutely no rights to the resources and that their continued use of the forest is criminal. Given that in some cases residents adjacent to protected areas (or their ancestors) have used the resources for centuries, some may argue that the criminalization of these traditional uses is unethical. Others may claim that the efficient point (with or without compensation) should be rejected on ethical grounds because it involves the coercion of a minority by the majority.

In a political-economic context, protecting land generates money for the host government from multilateral and bilateral agencies, from international nongovernmental organizations, and from tourists. Protected areas are often sited in politically and economically marginal areas, where not compensating residents typically costs the government very little (e.g., in votes and tax revenues). Mandating compensation may provide a disincentive to discriminate against politically weak or disfavored groups. If the government is required to pay compensation equal to the full value of the land, it may be dissuaded from taking an excessive amount of land from politically weak groups of individuals. In any case, compensation should only be paid where resident people have a legitimate claim to the land; compensation paid to recent squatters would only encourage encroachment into areas under consideration for parks.

A final point can be made based on the observation by Knetsch (1983) that whether an externality is considered negative or positive depends on the definition of rights. In the discussion of efficiency, we did not attempt to indicate the initial allocation of property rights to the resource because this allocation does not impact efficiency. (If there is a large disparity between the measures of WTP for a change and willingness to accept (WTA) foregoing a change for either residents or nonresidents, then the initial allocation of property rights will affect the efficient allocation of protected land. The potential for the existence of this disparity is discussed more fully later in this chapter.) We implicitly assumed that each group had a right to use a unit of land as long as its WTP for that unit is greater than the WTP of the other group. Thus, the loss of biodiversity beyond the socially optimal level is considered a negative externality. It could be argued, however, that nonresidents have no rights to flows of biodiversity protection. Under this view, when residents conserve biodiversity, they provide a positive externality to nonresidents. If nonresidents wish to consume this positive externality, they must pay (compensate) the residents for it.

## Conflict Mitigation

Even if one does not wish to compensate residents on the basis of efficiency, equity, or ethics, compensation may be justified on purely practical management grounds. In many regions of the world, people living closest to the protected biological resources will largely determine how many of those resources will survive. If the costs to residents of establishing a protected area are relatively high and the residents are not compensated for these costs, the magnitude of the conflicts between residents and managers of the protected area could potentially be quite large. Numerous studies throughout the world have shown that conflicts with residents significantly increase the costs of managing protected areas (e.g., West and Brechin, 1991). If these increased costs cannot be defrayed, the protected area will have a difficult future.

Several economists and policy analysts have shown that compensation may be a good political strategy for reducing conflicts over the siting of undesirable public enterprises such as hazardous waste dumps (e.g., O'Hare, 1977; Cordes and Weisbrod, 1979; Newberry, 1980; Sullivan, 1992). In the case of protected areas, it may be far less expensive to compensate residents than to battle them for control of an area's resources, if compensation leads to the relinquishing of claims and prevents further encroachment. Although enforcement has been shown to be effective in some cases (e.g., Kruger National Park, South Africa), in protected areas with large numbers of residents living nearby, the costs of effective exclusion through enforcement alone can be quite high. For example, despite the presence of numerous armed guards around the Royal Chitwan National Park in Nepal, Sharma and Shaw (1993) have demonstrated that illegal use of the park's resources was still prevalent. The administrative costs of enforcing the "efficient" allocation of habitat protection may be extremely high, such that compensating residents would actually lead to greater efficiency. Thus, compensation can be viewed as a means of minimizing enforcement costs for maintaining the protected status of parks and reserves.

Moreover, one could argue that acknowledging local rights can be an important mechanism for creating mutual respect between residents and outsiders. Mutual respect is important for opening constructive dialogues between residents and managers of protected areas. By treating residents as criminals, managers may only make them angry, defensive, and unwilling to cooperate. However, one must be careful to ensure that compensation schemes do not create incentives for further encroachment into protected areas.

There exist few examples of actual compensation payments to residents adjacent to protected areas. Most of the examples in the literature are anecdotal descriptions of partial compensation schemes. One of the more descriptive examples comes from Amboseli National Park in Kenya. Western (1982) reports that the Maasai people living around the park were granted a number of compensation arrangements to reduce their antagonism toward the park and the migration of wildlife through their lands during the wet season. In the mid-1970s, the Maasai were promised alternative sources of water and fodder, grazing fees, annual fees to cover losses from wildlife depredations, and mechanisms for deriving more benefits from tourism and the culling of wildlife. Some Maasai also received social benefits, such as schools, which were a byproduct of the development of the park's infrastructure. In just five years, there was a noticeable improvement in wildlife numbers and distribution and a decrease in the numbers of animals poached. Since that time, however, government cutbacks have ended the compensation fees and maintenance of alternative water sources (Wells and Brandon, 1992), and resident-park conflicts continue to threaten the park's viability (*New York Times*, 1991).

Although compensation may be a necessary condition for conflict mitigation, it is unlikely to be a sufficient condition. (We examine the reasons for this later in this chapter.) Moreover, even if compensation is deemed appropriate based on the criteria above, it may still be subject to problems associated with implementation in the field.

## PRACTICAL ASPECTS OF COMPENSATION

In addition to determining whether compensation is appropriate, governments face a number of practical aspects that must be dealt with in order to compensate residents appropriately. Indeed, problems associated with any of these aspects may indicate that an explicit compensation program is not the best approach to mitigating the costs incurred by residents and gaining their support for conservation activities. An alternative approach is outlined later in this chapter. The practical aspects that a government faces include strategic behavior, estimating the value of compensation, selecting recipients for compensation, selecting the form of compensation, and generating local conservation support.

### Strategic Behavior

The presence or absence of compensation may result in the development of undesirable strategic behavior by residents. Baumol and Oates (1988), referring to victims of "efficient levels" of pollution, argue against compensation on the grounds that it creates strategic behavior by residents that results in inefficiency. They maintain that "victims typically have available to them a variety of responses to reduce the damages they suffer. . . . [C]ompensation of victims is not economically efficient because it weakens or destroys entirely the incentive to engage in the appropriate levels of such defensive activities" (p. 24). In addition, in many areas where wildlife damage to crops, livestock, and human lives is a problem, the guarantee of full compensation may entice residents to engage in activities that would never have been considered optimal in the absence of assured compensation. Compensation may also reduce the incentives that residents have for engaging in certain behavior changes that may be part of the optimal solution. For example, while it may be more efficient for some residents to emigrate from the area or to intensify production on their current land rather than expand production to new land, compensation may diminish the incentives for such behavior.

Compensation of victims may lead to an increase in immigration to the peripheral zone of the protected area, which may lead to a socially excessive and ecologically damaging amount of economic activity in the peripheral zone. Such immigration also has the potential to increase the costs of compensation to levels far greater than originally anticipated. The incentives for immigration could be diminished if compensation is paid by a lump-sum payment to each household currently living in the delineated peripheral zone, or if all residents are registered and only they and their descendants are eligible to receive compensation benefits over time. However, if compensation is largely made through the provision of diffuse social services (e.g., schools, health clinics, technical assistance), it may be difficult to exclude recent immigrants. Some forms of compensation (e.g., wage employment, road building), especially those that tend to generate income, may lead to more migration to an area than do other forms (e.g., building schools, new techniques for weeding rice paddies). Thus, the effect that the form of compensation will have on migration must be considered. The problem of the protected zone as a magnet for immigration may warrant an exhaustive census of the zone's residents by the protected area managers. In this way, it will be easier for the managers to restrict benefits to current residents and their descendants. A few migrants may still arrive, but they will be relatively easy to locate and manage.

If the government permits the compensation package to be set by negotiation because, for example, it does not know the true value of the costs, there may be an additional incentive for strategic behavior. If each household had to sell its "share" of rights to the government, there may be holdouts who will demand compensation above the true value of their foregone benefits. If negotiation is done on a household-by-household or village-by-village basis, holdouts are likely to be few. But if negotiation is done on a collective basis, with a small number of representatives for a large number of residents, the residents may be able to act as a cartel and force the government to pay more than the actual costs.

There may be another way in which residents can increase the amount of compensation above the value of the true costs. As noted earlier, offering compensation implicitly recognizes the right of residents to exploit the biological resources as they wish. Some may claim that, by offering compensation, the government is, in a sense, rewarding socially undesirable activities ("social" in the sense of the national or global society). Although many authors point to the power difference between the residents and national and international interests as an important source of conflict (e.g., Hough, 1988), it is rarely recognized that the residents have a major source of power—in that the fate of the protected area lies in their hands. If residents recognize this power and realize that they are being "paid off" to not jeopardize the protected area's goals, they may attempt to force the government to pay excess compensation (potentially higher than the value of the global benefits) by threatening to undertake activities detrimental to the protected area, including activities that they normally would not consider profitable.

On the other hand, the lack of compensation may generate undesirable strategic behavior on the part of the residents. In order to establish protected areas, governments often take large portions of land to which residents believe they had rights. Having seen this action, residents may fear that more land will be taken in the future. If residents believe they will be undercompensated in the event of a taking, there may well be an increase in the rate of species depletion to make the land less suitable for biodiversity protection. For example, Ferraro (1994) notes that when the original boundaries of the Ranomafana National Park in southeastern Madagascar were walked by a team of U.S. scientists and Malagasy foresters, a relationship was made clear to many residents—heavily disturbed, deforested land was not considered to be desirable park land. A 1993 revision of park boundaries further reinforced this relationship by allowing many residents to keep cleared land out of the new boundaries. Some residents have thus begun to deforest peripheral-zone forests at an even higher rate in order to reduce the likelihood that the lands will be taken in the future.

By providing households or communities title to the land in the peripheral zone, managers of protected areas may reduce the residents' subjective probability attached to expropriation. The provision of land title may also provide the added benefit of helping current communities to resist any future waves of immigration that may be triggered by the benefits generated by the protected area management program. However, given the general distrust that residents often demonstrate toward the government and the lack of a "just compensation" clause in the legal code of many developing countries, the provision of land titles alone may not significantly reduce residents' fears of expropriation.

To reduce residents' fears, the government could guarantee compensation at the level of the full private—that is, financial—value of land in its optimal use. In order to assure the residents that compensation will in fact be forthcoming, the government could enter into an agreement with residents that would establish that if the government wishes to take more land, it will pay a specified dollar amount to the owners. If residents do not trust the government to honor its agreement, however, the threat of this strategic behavior persists.

## Estimating the Value of Compensation

If the government, after considering the potential strategic behavior in the presence and absence of compensation, still believes that compensation may be appropriate, then the proper value of the compensation must be estimated.

Very little detailed research on the local impact of establishing protected areas has been conducted, particularly in developing countries (Dixon and Sherman, 1990; Ghimire, 1991; Geisler, 1993). There has been even less research done on quantifying these impacts at a level that would permit some estimate of appropriate compensation (Western, 1982; Ruitenbeek, 1992; Ferraro, 1994; Kramer et al., 1994). Ferraro (1994) highlights how difficult it is to estimate the value of the costs incurred by residents, particularly in areas only partially integrated with the national economy.

One of the more technical problems, but one that is very important, is the estimation of economic impacts based on the residents' WTP to prevent the protected area, or on their WTA the protected area. Space limitations preclude a full discussion of this issue, but the analyst's choice of method can have a significant effect on the estimated level of appropriate compensation. Although some economists assert that in most cases the differences between the two measures are small or are the result of poor survey design (e.g., Randall, 1987; Arrow et al., 1993), the results of several empirical analyses point to large differences between the alternative value measures. Knetsch (1990) and Meyer (1979) together cite 15 studies that showed hypothetical WTA values that were 1.4–20 times larger than hypothetical WTP values. Although some economists state that WTP is the most appropriate measure (e.g., Russell, 1982), neither measure has been demonstrated to be superior. The existence of this disparity, however, indicates that assessments of losses as perceived by the residents may be seriously understated if WTP measures are used, and thus compensation payments based on such measures may not fully offset losses in the well-being of residents.

Ferraro (1994) and Kramer et al. (1994) provide frameworks for evaluating the economic costs to residents near protected areas. The WTP of residents can be approximated indirectly through the use of market and "shadow" prices (Ferraro, 1994; Kramer et al., 1994), or directly by using contingent-valuation techniques to query residents. The WTA of residents can be approximated directly through the use of contingent-valuation techniques (Kramer et al., 1994), or indirectly by convening "focus groups" that include representatives of the residents and of protected area managers. The most appropriate measure will depend upon the ultimate use of the results, time and money constraints, and the particular region in which the protected area is located.

The relative weights attached to each criterion used to evaluate the necessity of compensation will also affect the final value of the compensation package. For example, while concerns over equity may require that residents be compensated for all costs they have incurred as a result of the protected area, overall social well-being (efficiency) may require a limitation of compensation for substantial losses. Ultimately, the final level of compensation is a political decision that will have to be derived through dialogue between residents and managers of protected areas. The economic estimates can provide bounds or starting points for the bargaining process.

## Selecting Recipients of Compensation

Determining who shall receive compensation is often a difficult decision, and the choice depends on what criteria are used for determining the necessity of compensation. If conflict mitigation is the primary goal, only those residents deemed to be a threat to the goals of the protected area ought to be compensated. If compensation is driven by equity and ethical concerns, all affected residents ought to be compensated. There may be substantial numbers of residents who fall just over the line (physical or otherwise) dividing recipients from nonrecipients. If so, what types of conflicts will this generate? Should future generations be compensated? How will the recipient be defined, as an individual, a household, a village, or some larger unit? Given the great fluidity in the composition of many of these units even over short periods of time (e.g., households), some units may be easier to define than others.

In addition to these issues, results in Ferraro (1994), Kramer et al. (1994), and Shyamsundar (1993) indicate that costs to residents living near protected areas vary considerably. To pay each household the value of its losses, it would be necessary to know each household's costs, but making such a rigorous assessment on a household-by-household basis would be prohibitively expensive. Some method must therefore be used to approximate the losses, and the manner by which the approximations are made can have great impact. One concern is that if the loss to the average household is used as the standard payment per household, a significant number of people are going to be undercompensated and thus not satisfied, while others will be overcompensated, which may not be the most effective use of scarce conservation funds. Payment schemes in which a significant share of the population is undercompensated, but at least partially compensated, may appear to be "second-best" solutions. Since real-world policy typically relies on second-best solutions, however, such a compensation package should not necessarily be rejected. If compensation is provided to the village or other appropriate social unit, then many of the difficulties of sorting out individual household shares could be avoided by project managers.

## Selecting the Form of Compensation

Once the recipients of compensation have been defined, the managers (and residents) must then consider the form in which the compensation will be delivered. There is a wide range of choices—compensation could be delivered as cash, alternative land or resources, new technologies, employment opportunities, social services, infrastructure development, or a combination of several forms. In order to choose the most appropriate form, several questions need to be answered. Would residents be allowed to decide on the form of compensation according to their values and their preferences? Since each compensation package may have its own negative impacts on the resident population (e.g., dependency on income transfers), how should the project deal with these impacts?

To what extent is each possible form of compensation a substitute for the lost resources of the protected area? For example, if residents have lost the means to stay above a minimum income requirement, it is hard to imagine how a school or a health clinic, while perhaps desired by residents, will substitute adequately for lost resources. On the other hand, if the value of the lost resources is well established, it could be argued that cash compensation would more than substitute for the lost resources because cash has the added benefit of being far more liquid and mobile than natural resource assets. Moreover, if residents are risk-averse, they would rather have cash equal to the expected value of the foregone activities because the cash is certain. In rural areas of many developing countries, however, markets are highly imperfect, and thus residents may not be able to transform cash into the resources they need, or may be able to do so only at much higher prices than anticipated.

As with the estimation of the value of the compensation package, the form of compensation should be ultimately determined through negotiations between protected area managers and residents. If the government decides to pay compensation to residents, it may be appropriate to answer a final question, especially if one of the criteria used for evaluating compensation is "conflict mitigation": will compensation help achieve the conservation goals of the protected area?

## Generating Local Conservation Support

To the degree that it reduces conflicts between the management of protected areas and the people living nearby, compensation will have a positive effect on conservation. There is always the problem, however, that the residents may not be satisfied with mere compensation, given the magnitude of the benefits that they perceive outsiders to be receiving. Residents want to maximize their well-being. If it is difficult to link compensation to not damaging the protected area, residents may feel justified in asking for more than full compensation. If their demands are not met, they may take the compensation offered and continue using the resources as they did in the past.

Linking compensation to the promotion of resident behavior conducive to the goals of the protected area may be difficult, for several reasons. Such a linkage would require the government to offer compensation benefits spread over time and provided directly to each household. In this way, the benefits could be cut off if the household did not abide by its agreement to give up its claim to the protected area's resources. If a lump-sum initial payment (or extension of technology) is made or compensation is paid to large groups of households (in cash, schools, clinics, and the like), it would be difficult to exclude particular households. One could attempt to create a self-enforcing scenario by dictating that the entire community be cut off if one of its members violated the compensation agreement. But the ability of a community to penalize its members would vary across and within regions. Moreover, if the government did cut off an entire community's benefits, the same problem that necessitated compensation in the first place would continue to exist—that is, resource-hungry residents attempting to secure their future by exploiting the protected area's resources. If only a small number of households or villages do not abide by the agreement, it may be relatively easy (although perhaps not ethical) for the government to repress these households or villages with force, but there is no reason to automatically believe that this number would be small.

Moreover, many households are extremely poor. Compensating them for their foregone benefits would still leave them poor. If poverty is the primary factor driving them to exploit the protected resources illegally, compensating them is not likely to reduce their demand for these resources. On the other hand, although the underlying poverty-related roots of resource degradation would not be removed through compensation, not compensating the residents would only exacerbate their need to degrade the resources.

Thus, compensating residents will not necessarily generate local support for conservation endeavors, without which the protected area's future will always be unstable. Nor is compensation likely to change resource-use patterns in the peripheral zone in a significantly positive way. It may only be a matter of time before these resources are depleted and residents begin to demand the use of resources within the protected area. Even if development activities in the peripheral zone are able to increase incomes significantly, the result may be the same. Given current population growth rates and the residents' own desires to improve their economic well-being, residents will likely clear most of the peripheral zone within a few generations if their development is not somehow linked to protected area conservation. In the absence of local support for conservation, the residents will have little reason not to begin cutting down the forest within the protected area.

## SUMMARY OF POSITIVE AND NEGATIVE ATTRIBUTES OF COMPENSATION

This discussion indicates that there are some good arguments for compensating local residents, but there are some equally strong arguments against doing so. The various aspects of compensation are summarized in Table 9-1.

The relative strength of compensation as a strategy for safeguarding protected areas will be evaluated according to the weights that decision makers attach to the various aspects listed in Table 9-1. In some cases, it would appear that the benefits of compensation outweigh the costs, particularly if conflict mitigation is an important criterion for managers of protected areas. In other cases, however, it may be concluded that the costs of a compensation approach (e.g., the possibility that pressure on the protected area will increase because of compensation) outweigh the benefits. Moreover, it may be that an explicit program of compensation is not the most cost-effective way to achieve particular positive aspects of compensation listed in Table 9-1.

***Table 9-1*** Summary of positive and negative aspects of compensation

| Positive aspects | Negative aspects | Implications for parks and protected areas |
|---|---|---|
| Mitigation of inequity | Encouragement of undesirable strategic behavior | Limited potential for generating local conservation support |
| Mitigation of conflict | Significant complexity in estimating true costs and designing compensation program | |
| Mitigation of certain strategic behavior | Strong potential to reduce efficiency or to raise the cost of protection | Limited potential for changing the underlying factors contributing to the degradation of resources |
| Potential mitigation of discrimination against less powerful groups | Implied recognition of resident rights to use protected resources | |

Ideally, managers of protected areas would implement an approach that embodies the positive characteristics of compensation and avoids the negative, while accomplishing the conservation goals set for the protected area. It would seem that if such an approach existed, the choice of compensating residents or not compensating residents would be a moot one. In the next section, we argue that an explicit focus on modifying the incentive structure facing rural households living in the vicinity of protected areas can be such an approach.

## TOWARD A MORE EFFECTIVE APPROACH: THE USE OF ECONOMIC INCENTIVES

Anecdotal cases throughout the world give strong indications that if resident populations are opposed to the conservation goals of a protected area, the job of protecting the area's ecosystems is made considerably more difficult, if not impossible. It is known that residents attempt to maximize their well-being. They are guided by their preferences (material and nonmaterial) and are constrained by their available natural resources, labor, capital, and knowledge, as well as by the sociopolitical environment. Managers of protected areas and other collaborators may be able modify these preferences or constraints in order to link the conservation of resources within and outside protected areas to the maximization of resident well-being. At the very least, it may be possible to affect resident behavior so that the protected area's conservation objectives are not perceived by residents to be significantly impeding the maximization of their well-being. The modification of resident behavior can be achieved through a package of economic incentives—both positive and negative—that can influence the well-being of residents. These incentives might include government interventions such as subsidies, expanded opportunities for education, and law enforcement.

Although many proponents of conservation appear to be uncomfortable with or skeptical of the use of economic incentives, they often fail to recognize that economic incentives are driving the degradation in and around protected areas. Thus, in order to protect these areas, it will be necessary to alter these incentives in ways that promote conservation goals.

### Affecting Resident Behavior

Broadly speaking, in order to affect resident behavior to promote conservation, it is necessary to encourage households to reduce the amount of labor, capital, and natural resources that they allocate to activities that threaten the conservation goals of the protected area. It is preferable that residents reallocate these inputs to activities that do not reduce the biodiversity in protected areas or peripheral zones.

In general, there are five ways to promote such reallocations (Ferraro and Kramer, 1995): (1) compete for the labor currently allocated to destructive activities, (2) compete for the capital currently allocated to destructive activities, (3) compete for the biological resources currently being depleted to nonrenewable levels, (4) increase the information available to residents, and (5) encourage residents to adopt proconservation preferences. Since capital is usually not a large input into rural household production in the tropical developing countries, we will focus on the other four mechanisms.

In order to compete for the labor currently allocated to destructive activities, protected area projects can attempt to increase the opportunity costs of investing labor in destructive activities. By making production on lands already in use more profitable, or by creating new economic activities that do not depend upon the destructive use of biodiversity, protected area project personnel can encourage residents to reallocate labor away from destructive activities and toward nondestructive activities.

A protected area project can draw labor away from destructive activities in several ways. First, it can make labor more productive in activities that do not substantially deplete biodiversity. Labor productivity can be increased by introducing new labor techniques or complements to labor (e.g., affordable fertilizer) that improve labor productivity, by introducing entirely new production activities in which residents could more profitably invest some of their labor, and by improving markets and infrastructure in ways that make the output from desirable activities more profitable than the output from undesirable activities.

Second, a protected area project can reduce household discount rates by improving access to competitive credit markets or by increasing income. (The discount rate is the numerical way for comparing current and future costs. It is generally considered to be the premium that individuals are willing to accept for substituting present consumption for larger consumption in the future.) A decrease in the household discount rate makes it more profitable for a household to invest its labor in activities that will produce benefits in the future, as do many conservation-friendly activities. Third, a protected area project can increase residents' demand for leisure by increasing income. Fourth, the project can improve education services in the region. Residents who are better educated can take advantage of other employment opportunities, including those found in urban areas away from the threatened biological resources.

Finally, a project can use force as a negative incentive to prevent households from engaging in certain activities. Enforcement essentially makes investing labor in alternative activities that are legal more profitable than investing labor in illegal activities. The use of enforcement as part of a portfolio of incentives has received little attention during the past decade, partly because negative incentives were used almost exclusively during previous decades, with little success and considerable controversy (see chapter 1). Protection based on purely positive incentives is the ideal. But given the many constraints on devising effective positive incentives that achieve conservation objectives (as described below; also see Ferraro and Kramer, 1995), it is probably impossible to devise a system of protection in most areas based solely on positive economic incentives. This does not mean that reducing a protected area's dependence on elaborate enforcement activities is not a desirable objective. But the more a protected area can depend on self-enforcing conservation-friendly behavior by the resident population, the more likely will biological resources be protected over the long term.

When attempting to encourage residents to reallocate labor away from undesirable activities and toward desirable activities, two important questions should be asked: which labor is the proposed intervention competing for, and will the intervention absorb this labor? Rural residents of developing countries typically engage in a wide variety of economic activities throughout the year, each with its own particular requirements in terms of timing and the sex, age, and productivity of the workers. Such requirements placed on labor inputs, combined with the existence of a labor market, make a strategy based on labor competition difficult. Simply because activity A produces higher returns to labor (all other returns being equal) than activity B does not mean residents will reduce their allocation of labor to activity B. For example, suppose the goal is to reduce the amount of labor allocated to swidden (formerly called slash-and-burn) agriculture, which demands labor in April, July, October, and December. Introducing a more profitable activity that requires labor during other months will not compete for labor being devoted to swidden agriculture.

In order to compete effectively for labor, the managers of protected areas must have exceptionally good knowledge of the labor calendar, of the possible ways that households can alter this calendar, and of the sex-specific and age-specific aspects of labor allocation. The managers also must understand which activities, among all those practiced during a particular period, are the least profitable and which are the most profitable in terms of labor investment (an introduced activity may absorb labor away from a desirable activity rather than an undesirable activity).

Another strategy that protected area projects can implement to promote conservation is the introduction of alternative uses of threatened biodiversity—that is, uses that raise the opportunity costs of depleting the resources to nonrenewable levels. Projects can implement this strategy in several ways. They can increase the net benefits derived from the use of biodiversity by transferring new technologies, or by improving markets, prices, or infrastructure. They can actively aid the discovery of unexploited but potentially valuable biological resources. They can facilitate the participation of the resident population in the benefits to be derived from tourism or other nonconsumptive uses (e.g., biodiversity prospecting). All of these initiatives can help to increase the benefits derived from the nondestructive use of biodiversity.

In order for this strategy to be successful, however, several complementary conditions must be present. Residents must have secure rights to the benefits that result from not engaging in destructive activities. The economically profitable rate of use must be a biologically sustainable one. And for some activities, local institutions must be capable of coordinating the behavior of multiple households.

Protected area projects can also affect household behavior to promote conservation by altering residents' preferences or by offering residents new information that may encourage them to allocate more resources to conservation. Projects can achieve these two objectives through conservation education or the promotion of goodwill.

The ultimate goal of economic incentives is to make investments of labor, capital, and biological resources in desirable economic activities and investments in undesirable activities mutually exclusive. In other words, residents must be faced with a choice—they can either allocate their resources to the desirable activities and make $X, or they can allocate them to the undesirable activities and make $Y. They cannot do both sets of activities. If $X$ is greater than $Y$, the level of economic activities that threaten biodiversity will decrease.

## Advantages of Economic Incentives

When applied correctly, positive economic incentives (those that are actively endorsed by residents) ensure that the protected area's establishment does not negatively affect vertical equity, since, by definition, residents would have to be at least as well off as they were before the protected area was established. If residents are better off after the application of economic incentives, many of the ethical concerns can also be mitigated. Positive economic incentives also reduce conflict, since residents would by their own choice prefer to engage in alternative activities rather than in their current undesirable activities.

The use of incentives, both positive and negative, also reduces the probability of much of the strategic behavior described earlier. Since residents are not being explicitly compensated for each of their losses, there is no incentive to engage in risky activities that would never have been optimal in the absence of assured compensation. Although an increase in economic opportunities will undoubtedly attract immigrants, the positive incentives, in the presence of secure rights to benefits, ensure that most immigrants are likely to adopt the more profitable, proconservation activities. Moreover, the immigration that is often stimulated by the absence of clear property rights in a region will be discouraged. The use of enforcement as a negative incentive can also be used to discourage immigrants from engaging in undesirable activities. Furthermore, attempts by residents to force the government to pay excess compensation will be short-lived, since residents will soon see that if they do not adopt the alternative activities, they will be worse off.

The use of economic incentives also reduces the problems associated with identifying the recipients of compensation, since properly applied incentives should make most residents better off than they were previously. Moreover, the use of incentives can reduce the burden of estimating the costs to residents as a result of the protected area's establishment. Although it is helpful to have a good idea of the magnitude of these costs when considering plans for protected area projects, detailed information on costs and their distribution is not needed. However, in order to design an effective incentive package, it is necessary to clearly understand resident preferences and constraints. Similarly, although it is no longer necessary to choose the appropriate form of compensation, the choice of the appropriate form of incentives may be much more difficult.

Finally, the successful application of economic incentives ensures that the conservation objectives of the protected area will be achieved, because the incentives will make desirable and undesirable activities mutually exclusive. Using incentives to control the use of peripheral-zone resources can similarly ensure that biodiversity in these areas is not reduced to levels that cannot meet the needs of both the residents and the protected area.

## Finding the Linkages

The paucity of good examples from the field that demonstrate the ways in which economic incentives can promote conservation suggests that the correct application of incentives is not easy. The inability of protected area projects to promote the desired behavioral changes that will reduce pressure on biological resources is largely a result of the lack of understanding of how households interact with natural resources and of how one can affect household behavior in the desired ways. Ferraro and Kramer (1995) demonstrate in detail how a more precise conceptualization of household behavior can help the designers of protected area projects to identify more effective incentive packages. However, this research also indicates that there are many potential pitfalls. Protected area projects must not only be based on a clear understanding of how residents use resources (a difficult task) but also on how sociocultural aspects of production, imperfect markets, and government policies can affect the current incentive structure. Moreover, project designers and managers must clearly understand the ways that households can combine inputs (e.g., labor and money) to produce outputs (e.g., crops and forest products). Without this understanding, projects that create incentives for desirable activities may simultaneously increase the incentives for undesirable activities even further. Finally, those involved with protected area projects should be aware of the potential interactions among households, so that they can ensure that the identified threats are not simply displaced to other locations or times.

Despite these difficulties and the current lack of examples demonstrating the efficacy of economic incentives, the use of economic incentives should not be rejected. As we noted, it is economic incentives that are largely driving the current degradation of tropical ecosystems, and thus it is necessary to affect those incentives to halt the degradation. In order to improve the application of economic incentives in the field, we propose that protected area project personnel, government employees, scientists, and representatives of donor organizations do the following:

- Elaborate current hypotheses and assumptions about linkages between protected area projects and household behavior in order to capture more fully the actual circumstances in the field. Unless very precise conceptual linkages can be made between project interventions and the decision-making process used in households, the success of an economic incentive package will be highly improbable.
- Undertake quantitative analysis to further clarify how households may react to proposed protected area projects (see Ferraro and Kramer, 1995). There is a glaring lack of quantitative information that project participants can use to clarify their hypothesized linkages.
- Isolate the key aspects of various approaches that have ambiguous effects (e.g., income increases) and attempt to create an incentive package that reinforces the positive aspects of the interventions and mitigates the negative ones.
- Recognize that strong enforcement of protected area regulations must be brought back into the overall strategy in many regions. Enforcement should be viewed as part of a comprehensive package of positive and negative incentives.

In conclusion, if it is possible to successfully implement a package of positive and negative incentives that makes residents better off as a result of the protected area's establishment, then an explicit compensation program is not necessary. In some cases, however, compensation may be the most cost-effective way to encourage residents to permit a protected area to exist. In other cases, the optimal strategy may consist of a mix of explicit compensation payments and economic incentives (e.g., one that ensures compensation for any future lands taken and that provides residents with incentives to conserve currently protected areas). Protected area project personnel should strongly consider the costs and benefits of each approach. The only way to improve understanding of how economic incentives and explicit compensation programs affect protected areas is to apply them in the field, after having carefully considered the points raised in this chapter, and then monitor and evaluate their impacts.

The ideal way to ensure the long-term integrity of a protected area is to make it in the residents' self-interest to be actively interested in the area's conservation. Note that the word "actively" implies that residents do not simply ignore the protected area because they have more profitable activities in which to engage their household resources. Rather, this means the residents actively have a stake in maintaining the protected area in ways that achieve the area's conservation objectives. Only certain types of economic incentives can create this behavior. Scientists and the managers and other workers in protected area projects have begun to identify a few of these special incentives (e.g., revenue sharing), but the linkages are still not very strong. Identifying more incentives that encourage residents to actively support protected areas, and extending the use of these incentives in both scope and intensity, should be priorities for future research. The future of protected areas in tropical rain forests may depend in no small measure on how much we can learn about using economic incentives, and ultimately on how well we apply the lessons learned on a worldwide scale.

## NOTES

We have benefited from discussions with Nick Salafsky, Barbara Dugelby, Carel van Schaik, Marie Lynn Miranda, Subhrendu Pattanayak, and Priya Shyamsundar.

## REFERENCES

Arrow, K., R. Solow, P. Portney, E. Leamer, R. Radner, and H. Schuman. 1993. Report of the NOAA panel on contingent valuation. *Federal Register* 58: 4601–4614.

Barnett, S. 1991. Adding up the costs and benefits of conserving bio-diversity. In *Development research: the environmental challenge*, J. T. Winpenny (ed.), 156–58. London: Overseas Development Institute.

Baumol, W. J., and W. E. Oates. 1988. *The theory of environmental policy*, 2d ed. Cambridge: Cambridge University Press.

Blume, L., D. L. Rubinfield, and P. Shapiro. 1984. The taking of land: when should compensation be paid? *Quarterly Journal of Economics* 99 (1): 71–92.

Burrows, P. 1991. Compensation for compulsory acquisition. *Land Economics* 67 (1): 49–63.

Cordes, J. J., and B. A. Weisbrod. 1979. Governmental behavior in response to compensation requirements. *Journal of Public Economics* 10 (February): 47–59.

Dixon, John, and Paul B. Sherman. 1990. *Economics of protected areas*. Washington, D.C.: Island Press.

Ervin, D. E., and J. B. Fitch. 1979. Evaluating alternative compensation and recapture techniques for expanded public control of land use. *Land Economics* 19 (1): 21–41.

Farber, D. A. 1992. Economic analysis and just compensation. *International Review of Law and Economics* 12 (2): 125–38.

Ferraro, P. 1994. Natural resource use in the southeastern rain forests of Madagascar and the local impact of establishing the Ranomafana National Park. Masters thesis, Duke University.

Ferraro, P., and R. Kramer. 1995. *A framework for affecting household behavior to promote biodiversity conservation*. Washington, D.C.: EPAT/Winrock International Environmental Alliance.

Geisler, C. C. 1993. Rethinking SIA: why ex ante research isn't enough. *Society and Natural Resources* 6:327–38.

Ghimire, K. B. 1991. *Parks and people: livelihood issues in national parks: management in Thailand and Madagascar.* Discussion paper. Geneva: United Nations Research Institute for Social Development.

Gillis, M. 1991. Economics, ecology, and ethics: mending the broken circle for tropical forests. In *Economics, ecology, and ethics: the broken circle*, F. H. Bormann and S. R. Kellert (eds.), 155–79. New Haven, Conn.: Yale University Press.

Haas, P. M., M. A. Levy, and E. A. Parson. 1992. Appraising the earth summit: how should we judge the UNCED's success? *Environment* 34 (8): 6–33.

Hodge, I. D. 1989. Compensation for nature conservation. *Environment and Planning* A 21:1027–36.

Holt, H. B. 1988. Property clause regulation off federal lands: an analysis, and possible application to Indian treaty rights. *Environmental Law* 19 (2): 293–320.

Hough, J. L. 1988. Obstacles to effective management of conflicts between national parks and surrounding human communities in developing countries. *Environmental Conservation* 15 (2): 129–36.

Johnson, M. B. 1977. Takings and the private market. In *Planning without prices*, B. H. Siegan (ed.), 63–111. Lexington, Mass.: Heath.

Knetsch, J. L. 1983. *Property rights and compensation: compulsory acquisition and other losses*. Toronto: Butterworth.

Knetsch, J. L. 1990. Environmental policy implications of disparities between willingness-to-pay and compensation demanded measures of values. *Journal of Environmental Economics and Management* 18:227–37.

Korsmo, H. 1991. Problems related to conservation of coniferous forest in Norway. *Environmental Conservation* 18 (3): 255–59.

Kramer, R. A., N. Sharma, P. Shyamsundar, and M. Munasinghe. 1994. *Cost and compensation issues in protecting tropical rainforests: case study of Madagascar.* Environment Working Paper No. 62. Washington, D.C.: World Bank, Environment Department, Africa Technical Department.

McNeely, J. 1991. Bio-diversity: the economics of conservation and management. In *Development research: the environmental challenge*, J. T. Winpenny (ed.), 145–55. London: Overseas Development Institute.

Meyer, P. A. 1979. Publicly vested values for fish and wildlife: criteria in economic welfare and interface with the law. *Land Economics* 55 (2): 223–35.

Miceli, T. J. 1991. Compensation for the taking of land under eminent domain. *Journal of Institutional and Theoretical Economics* 147 (2): 354–63.

Nash, N. C. 1993. Bolivia's rain forest falls to relentless exploiters. *New York Times*, 21 June, 1.

Newberry, D. M. G. 1980. Externalities: the theory of environmental policy. In *Public policy and the tax system*, G. A. Hughes and G. M. Heal (eds.), 106–49. London: Allen and Unwin.

*New York Times*. 1991. An African park in peril. 19 May, section 5, page 19.

O'Hare, M. 1977. Not on my block you don't: facility siting and the strategic importance of compensation. *Public Policy* 25:407–58.

Pollot, M. L. 1993. *Grand theft and petit larceny: property rights in America*. San Francisco: Pacific Research Institute for Public Policy.

Posner, R. 1986. *Economic analysis of law*. Boston: Little, Brown.
Randall, A. 1987. *Resource economics: an economic approach to natural resource and environmental policy*. New York: Wiley.
Ruitenbeek, H. J. 1992. The rainforest supply price: a tool for evaluating rainforest conservation expenditures. *Ecological Economics* 6:57–78.
Russell, C. S. 1982. Publicly vested values for fish and wildlife: comment. *Land Economics* 58:559–62.
Sharma, U. R., and W. W. Shaw. 1993. Role of Nepal's Royal Chitwan National Park in meeting the grazing and fodder needs of local people. *Environmental Conservation* 20 (2): 139–42.
Shyamsundar, Priya. 1993. Economic implications of tropical rain forest protection for local residents: the case of the Mantadia National Park in Madagascar. Ph.D. diss., Duke University, School of the Environment.
Speth, J. G. 1990. North-south compact for the environment. *Environment* 32 (5): 16–20, 40–43.
Sullivan, A. M. 1990. Victim compensation revisited: efficiency versus equity in the siting of noxious facilities. *Journal of Public Economics* 41 (2): 211–25.
Sullivan, A. M. 1992. Siting noxious facilities: a siting lottery with victim compensation. *Journal of Urban Economics* 31 (3): 360–74.
Tietenberg, T. 1988. *Environmental and natural resource economics*, 2d ed. Glenview, Ill.: Scott, Foresman.
Todd, E. C. 1976. *The law of expropriation and compensation in Canada*. Toronto: Carswell.
Ward, J. T. 1986. Compensation for Maori land rights: a case study of the Otago tenths. *New Zealand Economic Papers* 20 (2): 3–16.
Wells, M., and K. Brandon, with L. Hannah. 1992. *People and parks: linking protected area management with local communities*. Washington, D.C.: World Bank, World Wildlife Fund, and U.S. Agency for International Development.
West, P. C., and S. R. Brechin, eds. 1991. *Resident peoples and national parks: social dilemmas and strategies in international conservation*. Tucson: University of Arizona Press.
Western, D. 1982. Amboseli National Park: enlisting landowners to conserve migratory wildlife. *Ambio* 11 (5): 302–8.
Willis, K. G., J. F. Benson, and C. M. Saunders. 1988. The impact of agricultural policy on the costs of nature conservation. *Land Economics* 64 (2): 147–57.
Yapp, G. A. 1989. Wilderness in Kakadu National Park: aboriginal and other interests (Australia). *Natural Resources Journal* 29 (1): 171–84.

# 10

# Toward a New Protection Paradigm

*Carel P. van Schaik and Randall A. Kramer*

During the past century, the standard measure for safeguarding the maintenance of biodiversity has been the establishment of protected areas in which consumptive uses by humans are minimized. Over the years, the design of protected areas has evolved from the creation of small refuges for particular species to the protection of entire ecosystems that are large enough to maintain most if not all their component species and that are mutually interconnected wherever possible. While many other, equally important, measures are now being contemplated and implemented (e.g., comprehensive land-use planning, sustainable development), protected areas remain the cornerstone of all conservation strategies aimed at limiting the inevitable reduction of this planet's biodiversity (e.g., *World Conservation Strategy, Caring for the Earth, Global Biodiversity Strategy*).

Existing protected rain forest areas suffer from an array of problems that reduce their effectiveness in a broad conservation strategy. They cover a scant 5 percent of tropical rain forest habitats (WCMC, 1992)—arguably not enough to forestall species extinction, especially since the proportions of areas protected vary appreciably from region to region. Protected areas are often not sited appropriately, and they are often too small to maintain the full diversity of their communities. They will in future be affected by external forces (Neumann and Machlis, 1989), such as changes in local climates caused by extensive deforestation, pollution, or fires emanating from outside; introduced exotic species; and global climate change, which in parts of the tropics will likely manifest itself as an increase in the frequency of long droughts. Fortunately, these existing and anticipated threats are being addressed in some countries and regions by measures such as integrated land-use planning, redesigning parks, and establishing corridors, although ecologists are concerned that not enough is being done (see chapter 3).

These shortcomings of protected area networks are significant and need to be redressed, but human activities currently pose far more serious threats to protected areas. These activities include conversion of rain forests due to an expanding agricultural frontier; extraction of game, fish, or timber; use of the forest to graze cattle; fires deliberately set or spreading from agricultural land; and different forms of mining and infrastructural projects, such as roads and hydroelectric dams (see chapter 4). Indeed, some protected areas exist on paper only.

Although the seriousness of these pressures on protected areas is increasingly recognized, the current conservation paradigm emphasizes that a compromise between human needs and the needs of nature can be found and advocates sustainable use rather than strict exclusion. Yet protected areas are needed, not to satisfy some Western romantic ideal about paradisal nature unspoiled by humankind's uncouth hands but because a considerable number of species are vulnerable to extinction due to overexploitation or disturbance. As we argued in chapter 1, tropical rain forest organisms are particularly vulnerable, and examples abound of local species losses due to subsistence and commercial hunting (Redford, 1992) or selective logging (Thiollay, 1992; Gorchov, 1994). Theoretically, it may be possible to design harvesting schemes that are sustainable and have limited negative impact on the other species in the system, but in practice such harvesting is unlikely to be commercially attractive due to the low sustainable extraction rates. Ironically, markets create pressures to increase extraction rates as species get scarce and prices rise. Even subsistence use is not immune to the impact of increases in individual resource demands and in the human population. Finally, brief periods of unsustainable use may lead to local extinction, and in a fragmented habitat such losses are all but irreversible.

These risks should be enough to invoke the precautionary principle for biodiversity maintenance (Myers, 1993) and give high priority to setting aside a modest amount of land as strictly protected areas. The proper response to the difficulty of maintaining protected areas is not to formulate an alternative based on extractive uses but rather to find novel ways to defend these areas and to set aside new areas. This will no doubt be an uphill battle in an increasingly crowded world, where unexploited forests are diminishing and degraded lands are on the rise.

It is against this background that several contributors to this book have noted that the world is not demonstrating the collective resolve needed to maintain the existing protected areas. We have singled out tropical rain forests, but there is little reason to be more sanguine about the state of other biomes. The aim of this chapter is to address the causes of the perilous state of protected rain forest areas. Only when the causes of the threats to these areas are correctly identified is it possible to contemplate the design of effective solutions. Building on the previous chapters in this book, we begin by formulating some general principles that should underlie all attempts at managing protected areas. Next, we identify two sets of causes of degradation of protected areas that are mediated through two broad classes of actors: small players and big players. We then develop recommendations that follow from giving proper attention to biodiversity protection.

## BASIC PRINCIPLES FOR DESIGNING SOLUTIONS

It would be arrogant to presume that it is possible to present a definitive list of recommended courses of action that would be guaranteed to improve the state of protected areas in the tropical forest biome. As the saying goes, all conservation problems are the same, but each solution must be unique. Too many idiosyncratic factors foil any attempt at the design of generic rules that would rise above the trivial. There is no substitute for careful local analysis of the causes of park degradation (see chapter 5). However, we hope to identify the basic principles upon which to base a sound and lasting strategy for the maintenance of tropical forest biodiversity, taking the need for inviolate protected areas as the point of departure.

As noted by Kramer and Sharma in chapter 8, a conservation area produces many benefits, but the magnitude of these benefits, real or perceived, will vary with the distance from the area. Some of these benefits accrue only to the local community and decrease as one moves away from the protected area; one example is the area's ability to regulate local climates and water flows. Other benefits accrue mainly to the inhabitants of nearby cities (e.g., water supply, recreation opportunities) or the nation (e.g., reduction in natural disaster recovery expenditures). Yet others accrue to all world citizens, although they may be valued differently by different people. For instance, the carbon stored in an intact forest limits the greenhouse effect; this benefit would be especially appreciated by citizens of regions that would most strongly be affected by global warming (e.g., low-lying countries, or regions that are currently major grain producers). The same holds for biodiversity. The existence value of tropical rain forest species is often appreciated more in big cities or in distant (often temperate) lands than in or near the forest itself.

However, the presence of a protected area also entails economic costs, in particular opportunity costs, in that protection prevents other uses of the area. This opportunity cost (the cost of foregone alternatives) accrues most significantly to the local communities. It is difficult to assess how the balance of costs and benefits varies with distance from the protected area, but this balance will often be negative in the area directly adjacent to it. Of course, where the benefits are not (yet) valued highly, this will make the perceived balance all the more negative. As a result, local communities are less likely to accept the opportunity costs imposed upon them by the presence of a protected area, simply because others elsewhere benefit from this presence. Since conservation of biodiversity, like good government, should aim at serving the common good, this distributional imbalance leads to two fundamental conclusions, encased in the first two general principles.

## Principle One

The first general principle is that active protection—that is, law enforcement—should be a fact of life for protected areas. There may be a few exceptions to this principle, for instance, when a protected area sustains intensive agriculture or locally profitable ecotourism, provided that the communities are close-knit and capable of self-policing.

Many people are opposed to this principle. It is often said that education will change the attitudes of local communities toward protected areas and will make communities see the benefits produced by these areas. However, while education and extension have the virtue of eliminating many apparent conflicts, the real conflicts of interest do not go away. In many cases, exploitation of a protected area or its conversion would be the rational course of action for local communities; hence, it cannot be expected that local communities will invariably favor protection. Moreover, even where communities as a whole benefit from protection, support will not be unanimous: there will always be individuals who will benefit from exploitation. Finally, some of the extractors may not be local people. For instance, communities around U.S. national parks often benefit from the stream of tourists drawn to the park and the attractiveness of the region to high-technology industries, but nonetheless there are serious poaching problems in many of the parks because some species or their parts (e.g., gallbladders of bears) garner very high prices on the international market.

It is also often argued that active defense of protected areas can never work (West and Brechin, 1991). However, one could counter that it has rarely been tried in a situation where support and incentives were also offered at the same time. Where this combination has been offered, it is reported to work better than either support or enforcement alone (Hannah, quoted in McNeely, 1993).

Finally, it is often said, both in the nations of the North and South, that there is no moral justification for denying the local poor access to resources locked up in protected areas. This argument undoubtedly has some force, especially given the large social inequities within tropical forest countries and between them and other countries whose wealth is also largely based on the same destructive resource mining. This argument, however, pits the interests of the current generation of local people against those of the as yet unborn, and it also assumes that local processes are not driven by developments elsewhere (Fearnside, 1993). If the political will can be mustered, alternative solutions may be found, especially when external support is available. Indeed, these are the very situations in which external support is called for.

## Principle Two

The second principle also emerges from the analysis of the distribution of costs and benefits of biodiversity protection. It states that it is reasonable to ask that the beneficiaries elsewhere be prepared to somehow pay for these benefits, which they currently receive at no cost. Hence, we should be seeking methods of capital transfer to help with the establishment or rehabilitation of protected areas (see chapter 8). Devising the exact costing formulas, transfer processes, and performance monitoring will be challenging, but this approach is necessary to mobilize the funding required. For instance, to incorporate the geographic imbalance in existence value of biodiversity, one could request outside support for the additional actions required to turn a sustainable land-use plan into one that preserves the highest amount of biodiversity. This will likely mean soliciting help from developed nations. In addition, the beneficiaries increasingly include affluent city dwellers inside the countries containing the protected areas. If classic ways of capital transfer, such as taxation, are not in place, other avenues to support the protected areas may be found. This might include, for instance, support by private foundations.

## Principle Three

Effective and lasting solutions for conflicts of interest require the involvement of all stakeholders (see chapter 7). In the case of protected areas, these stakeholders include not only local communities and other nationals of the countries concerned but also the international community. This has two implications, again encased in two additional general principles.

The third principle is that foreign involvement in the management of a country's biodiversity is justified. At first sight, such involvement would seem to represent a violation of a country's sovereignty. However, actions in one country may cause problems in another country. Keohane et al. (1993) point out that there are two kinds of international environmental problems: transboundary and "commons" problems. Pollution and the diversion of rivers are examples of the first kind. In the case of biodiversity, the protection of long-range migratory species would fall into this category. Because countries can claim that their national interest is at stake, direct negotiation between the countries concerned is a natural approach to solving these issues. Commons problems are exemplified by the release of greenhouse gasses or chemicals that affect the tropospheric ozone layer. Here, broad international collaboration is called for. Biodiversity preservation in general would seem to fall largely within this category. International involvement here is predicated on the notion that just because a country happens to be the home of endangered species, that country does not have the right to let the species go extinct (see chapter 6). This idea also implies that all of humanity is responsible for maintaining biodiversity. Because different nations hold different views of this matter, however, clashes in cultural values become apparent. This situation is more similar to human rights than to, say, the ozone problem, where interests of different nations are more likely to coincide.

Many have argued that some form of international governance is needed to solve urgent international problems (e.g., Boza, 1993). The world is not yet ready for peacekeeping forces, however, let alone for uninvited "nature-keeping" forces. Thus, the best way in which the international community can get involved in the protection of biodiversity at the national level is through carefully negotiated binding international agreements (French, 1992).

### Principle Four

The need to involve all stakeholders directly or indirectly in the management of protected areas has a second implication. It has become commonplace to advocate the devolution, or transfer, of management authority to those most directly involved, namely, the local communities. In many conditions, as noted by Miranda and LaPalme in chapter 7, such devolution improves productivity and reduces the risk of overexploitation (see also Lynch, 1992). In protected areas, active involvement of local communities is likely to improve protection significantly (Wells and Brandon, 1992). Hence, active involvement of local communities in conservation is mandatory.

However, Miranda and LaPalme also noted that conservation of biodiversity is unusual in that local actions often have impacts far beyond the limits of the community's domain. Thus, the fourth principle states that devolution of protected area management is likely to be effective, but that the interests of the other stakeholders, be it the nation or the international community, should always be represented—if not at the executive level, then at least at the oversight level. Thus, the strong point of devolution (reduced bureaucracy) should be married to the strong point of state involvement (promotion of the common good). We urgently need models of how effective involvement of local communities can be achieved without depletion of protected areas.

## THE ACTORS BEHIND PROTECTED AREA DEGRADATION

As noted in chapter 4, we distinguish two sets of causes of protected area degradation mediated through two broad classes of actors: small players and big players. The small players are expanding local populations, indigenous people who have become integrated into a market-oriented economy, and people displaced from overcrowded rural or even urban areas. Land seekers and extractors flock to the frontier where existing land-use limitations are weakly enforced or not enforced at all. These people have in common that they have little economic power but are determined to eke out a living. In any individual case, it seems eminently reasonable to accommodate pressing local needs. However, since this is happening almost everywhere and at the same time, it leads to the tragedy of attrition, a steady erosion of park integrity. Attrition must be confronted by shoring up the defenses of protected rain forest areas in the face of increased pressure on their resources. In some densely populated places, such as West Africa, this pressure has already reached the level of a war for the last remaining pieces of wildland (e.g., Southgate and Clark, 1993).

The second class of actors consists of the big players: organized and well-connected elites, with or without explicit involvement of the government, or even the government itself. The actions of big players are not always illegal because the players either represent the law or have the power and influence to change the law. Sometimes their actions are genuinely aimed at improving the national economy or at regional development, and they have been encouraged by international agencies or bilateral aid programs. Where such projects conflict with existing protected areas, the conflicts have often been resolved by ignoring or changing the protected status of the land, usually because the responsible agency is overruled by politically more powerful departments. Mines, roads, hydroelectric dams, timber operations, and plantations are examples of potentially destructive projects frequently pursued by big players. Often, the development activity merely enriches the elite.

Sometimes the two sets of players interact. Large government or corporate projects often attract workers to the area, many of whom stay on to grow cash crops after the construction phase is completed. Most projects involve road construction. Roads in particular have caused incalculable harm to biodiversity in protected areas because they enable settlement and the timely transport to markets of perishable produce. Other projects displace people who are then forced to convert existing forest land (e.g., Brechin et al., 1994).

Despite frequent interactions, the distinction between small and big players is important for two reasons. First, most attempts to improve the fate of protected areas have been aimed at the first group of actors. In some cases, however, small players are not the major agents of park degradation, and in other cases they are merely a symptom of underlying policy failures (Fearnside, 1993). Second, to strengthen protected areas we need action directed at different levels: at the level of project and national policy in the case of the small players; at the level of national policy and international agreements and politics in the case of the big players.

## DEALING WITH THREATS POSED BY SMALL PLAYERS

Local problems with protected areas are a product of high pressure from the outside and low resistance on the inside. We want to address both, beginning with the mounting pressures. As noted by van Schaik et al.

in chapter 4, the increasing pressure on protected areas is a complex phenomenon, caused by various, interacting forces.

## Reducing Pressures

The pressures on protected areas arise in three different ways, and separate strategies will be required to effectively reduce them. First, there are local communities living at the periphery of or even inside protected areas. Pressures can be reduced by providing alternative sources of livelihood to those dependent on the forest's resources. Second, short-term stabilization of the locally generated pressures on protected areas is needed, but as long as human populations and aspirations continue to rise, this will not be enough. More radical reductions of pressure need to be considered by offering attractive alternative options elsewhere. Third, there are the disenfranchised who flock to the frontier in desperation, having no other options in the market economy than to become subsistence farmers, extractors, or miners. Only changes in government policies can effect improvement in this category of encroachers.

### *Local Communities: The Short-Term Perspective*

Governments of civilized nations have the duty to ask their citizens to accept restraints on their freedom of action when it serves the common good. And governments have established enforcement mechanisms in implicit or explicit recognition of the underlying conflict of interest. Traffic and zoning regulations and their enforcement are good examples. In the case of tropical forest parks, governments can claim forest lands as national property because they serve national and international interests.

In practice, of course, local communities often show strong hostility toward protected areas. Ultimately, the hostility reflects a lack of acceptance of the limitations imposed by the government. One of the reasons for this is historical: the developing nations in the tropics have nationalized some 80 percent of forest lands (Panayotou and Ashton, 1992), often without regard to the traditional arrangements that preceded their authority or that of previous colonial masters (see chapter 7). Another reason, however, is that local communities do not always perceive the government as representing their interests. Thus, they are unwilling to accept the government's authority even where the people arrived on the scene well after the protected area had been established.

These problems are compounded if the local government shares the view of the populace. Since the local government cannot be held accountable by the international community—it is not a signatory to international agreements—this creates an acute conflict of interest with the national government, which will likely be resolved by ignoring the international commitments.

Despite these different manifestations, the hostility toward protected areas is essentially based on economics. Local people are loath to accept the opportunity cost imposed upon them by government. The solution, therefore, could also be an economic one, and it is primarily a matter of semantics whether one wishes to label such financial support as compensation or not (see chapter 9).

Some have argued that to improve protection, it is necessary to strive to remove the opportunity costs by making protected areas profitable. It is essential that locally perceived values of protected areas come to reflect reality as much as possible. This is not easy, because some of the benefits are expressed in saved expenses that are not visible unless we remove the services produced by the protected area and are therefore taken for granted (e.g., the watershed protection provided by intact forests). One way to change perceptions is for conservationists to travel with local leaders to other areas so they can gain first-hand experience with successes and failures of conservation. Nonetheless, it is unrealistic that changed perceptions alone can turn all protected areas into "win-win" solutions, where local communities consider that a protected area makes a positive contribution to the local economy.

One of the few activities that, if carefully controlled, can lead to lasting profitable use of a protected area without diminishing its value is ecotourism. Although rain forests are less likely to spawn the large-scale safari tourism that has turned savanna parks into revenue generators, there are local possibilities where spectacular scenery or large fauna (e.g., great apes) provide a unique experience for visitors. Revenues may support local communities by providing employment and business. In some of the current attempts to develop ecotourism, however, local people are often not sufficiently involved, because the tour operators or the conservation services provide all the local services, including both the accommodations and the guiding (Boo, 1990). Local support can only be sustained where tourism revenues are returned to local communities and the protected areas. (For a broader discussion of ecotourism see Boo, 1990; Brandon, 1993; Healy, 1994.)

Where protected areas cannot be self-supporting, support from the outside is appropriate, because the land-use conflict reflects an underlying conflict of interest between individuals and (parts of) local communities on the one hand and the regional, national, or international community on the other hand. Determining the amount and the form of outside support is resolved through negotiation between the local communities and the outsiders bringing the support—probably in most cases a combination of national and international players (see chapter 8). Although the outcome of this process cannot be predicted in detail, it should lead to the following predictable and dependable outcome: the parties agree that the contract implies both the recognition on the part of the government of the opportunity costs suffered by local communities and the recognition by the local community of the government's ownership of the protected area. This will help to ensure that communities will accept the government's law-enforcement activities (Wells and Brandon, 1992).

Recent analyses of several projects aimed at improving the integrity of protected areas by supporting the people living nearby have suggested some other lessons (see chapter 5). Action aimed at the periphery is not necessarily the most effective at stemming park degradation. Some projects aimed at the peripheral zones of protected areas, or buffer zones, were so successful that the area actually started to attract migrants who would benefit from the services provided in the buffer zone (Oates, 1995). From the perspective of the original objective, this may be considered a perverse impact of the conservation action. Such unwanted impacts are more likely when the conservation project has no control or management authority in the buffer zone.

The involvement of local communities through economic incentives should encourage those activities that are not dependent on protected areas. Alternative techniques of producing food crops, animal protein, and fiber can be encouraged through technical and financial assistance underwritten by protected area support funds. If support of these activities is guaranteed by a contract with local communities, and continuation is contingent on absence of encroachment, a recognized link between continued economic well-being and biodiversity protection can be developed. Note that this is distinct from the prevailing model of economic linkages to conservation that requires direct use of the protected resources. Here we advocate a linkage between conservation and development activities carried out in different parts of the landscape. This spatial linkage should be taken into account when the boundaries of project areas are defined.

### *Local Communities: The Long-Term Perspective*

Even in the absence of immigration, actions aimed at improving the economic position of local people may not be enough in the end. Increases in local population and per capita resource demands, and decreases in available land due to degradation or conversion to residential or industrial purposes, will still mean that pressures on protected areas continue to rise. Intensive use and high biodiversity are not compatible. Hence, encouraging more drastic processes that would lead to a permanent reduction of pressure should also be encouraged. The most obvious way to achieve this is to encourage developments elsewhere that would act as magnets, pulling people away from the economically less attractive peripheral regions.

Examining the land-use cascade presented in chapter 2 suggests several ways of doing this. First, the large and growing category of degraded land produces a demand for new cultivated land. Hence, a concerted effort to convert degraded land back into productive land should have the effect of reducing pressures on protected areas. Second, activities that will help to intensify agriculture on the existing agricultural land will slow the demand for new land, because it will do away with the need for fallowing and because it will increase yields per unit area in locations with permanent agriculture. This would be especially effective where land hunger rather than economic development or population growth drives the pressure on protected areas.

Finally, the most effective long-term solution is to provide aid aimed at improving urban infrastructure elsewhere that encourages industrial development. This development would act as a source of employment for the supernumerary rural poor, much in the same way that in the northern hemisphere the rural population surplus was absorbed into the developing cities. Industrialization and urbanization in the tropical world are proceeding apace with this historical trend, and should lead to significant reductions of pressure, if the options available to people are considered more attractive than subsistence farming or extraction. Moreover, increasing urbanization may lead to a considerable slowdown of total population growth, and to an increased demand for recreation in natural areas.

While these alternative approaches should provide more effective long-term protection for protected areas, the source of funding is not immediately clear. In some cases, it will be possible to develop a resettlement plan together with the communities living at the park's periphery, as part of the compensation package covering a protected area conservation project. But in other cases, support for development that attracts people from frontier areas should be independent from the compensation issue, and incorporated into regional planning that seeks the optimum mix of conservation and development priorities. Such integrated planning could be implemented with support from bilateral and multilateral agencies, provided they are willing to modify their financing arrangements and allow the mutual alignment ("bundling") of different conservation and development projects that ultimately serve the same biodiversity conservation objective.

### *Migration Policies*

Much of the tropical deforestation is driven by developments and policies originating far from the forest frontiers, including international ones (Southgate and Clark, 1993; Skole et al., 1994). Thus, the solutions have to be found partly in government policies that discourage migration into the frontier regions where protected areas are often found. This is another area in which foreign assistance—for instance, by the Global Environment Facility (GEF)—can be very effective (see Brown and Wyckoff-Baird, 1992). Examples of policies effective in reducing migration are the revoking of subsidies on pasture development, the discontinuation of government resettlement plans, and the cessation of road building into wilderness areas. In Brazil, the last few years have seen a considerable slowdown in deforestation after changes in government policies (Skole et al., 1994; but see Fearnside, 1993).

## Increasing Resistance

The success of all the activities suggested above is predicated on a simultaneous increase in the resistance of protected areas to untoward use or conversion. There are some obvious reasons for the inability to defend protected areas. One of them is that some governments lack authority because they are holding their citizens to standards different from those they adopt themselves. Another reason may be that foreign donors are not keen on strengthening law enforcement as part of development support. Yet another reason, ironically, may be the new emphasis on sustainable use advocated by major international conservation policy documents. Sustainable use of resources has received so much attention that conservation, to many people, has become synonymous with sustainable use. Thus, the emphasis on sustainable development has inadvertently contributed to a reduced commitment to protection of biodiversity.

Let us assume, however, that governments are convinced of the need for protected areas. What options do they have to improve the effectiveness of protection? In general, both increased funding and improved management are necessary. Where countries have limited resources to develop or maintain effective management institutions, help is needed.

Financial support may best be targeted directly at the protected area, provided that the responsible national government agency is supportive and that the regular enforcement authorities cooperate. Such direct targeting of financial assistance can be accomplished through debt-for-nature swaps and trust funds. Loans are not the right vehicle to support protected area management, since they are meant to support income-generating activities. Debt-for-nature swaps hold great promise for establishing or maintaining the limited amount of tropical forests placed in protected status. They have been used to establish new protected areas in at least 13 countries (Brown and Wyckoff-Baird, 1992). In some countries, debt-for-nature swaps have provided ongoing funding for protected area management activities such as salaries and logistics.

Trust funds are an attractive mechanism for support in perpetuity of protected area operations, community-based development activities, and maintenance of park infrastructure (see chapter 8). They can be funded through debt-for-nature swaps or through grants from the GEF, individual countries, or nongovernmental organizations (NGOs). By tying establishment of the trust funds to specific protected areas, it is possible to improve accountability and ensure the continuity of funding for park management, insulated from government deficits and changing political priorities. The best way to manage such trust funds is still subject to experimentation, and will probably vary locally.

However, protected area management itself can also be improved (see Box 3-1). It is bound to be inadequate when, as in quite a few countries, salaries of park staff are insufficient to maintain a family or arrive quite irregularly. Ironically, such salary problems sometimes mean that more park staff may make protection worse rather than better, because they almost guarantee local corruption. For instance, poaching by staff or poaching overlooked by staff in return for payment is common in many Southeast Asian countries, according to information presented at the Wildlife Conservation Society regional meeting in Bali, Indonesia, in August 1994. Apart from improving salaries, it may be possible to encourage staff to protect rather than exploit by providing some linkage between performance and reward. For instance, performance bonuses can be given at regular intervals and will depend on the degree to which the protected area's integrity is maintained. Creative new management models, involving NGOs (as in South America) or private-public partnerships, may also improve resistance, but there is little experience so far. Institutional structures and cultures also tend to make it difficult to work more closely with local communities. In sum, the management side of the resistance problem deserves more systematic analysis than it has received.

Finally, local powers opposing the park service's authority may often extract resources or convert the area by force or the threat of force, leaving the best park managers almost helpless. This situation is bound to become more common as countries devolve many tasks. Some (e.g., Susskind, 1994) have suggested that there is a role for the national military in this regard. This is not as farfetched as it sounds, since the role of the military is to protect the nation's interest, usually against outsiders but in case of emergency also against rebellious insiders. Moreover, the military is often the only power with authority and is the best organized and equipped institution in the country. Use of the military, however, may cause resentment among local residents and reduce local conservation support, so it should be considered only as a means of last resort.

A frequently overlooked benefit of scientific research is that researchers, foreign and domestic, have detailed information on the local situation, tend to be less vulnerable to intimidation by local powers, and tend to have the contacts at the central level to make sure that information reaches those who need it and can put it to effective use.

## DEALING WITH THREATS POSED BY BIG PLAYERS

Governments, corporations, or military organizations also undertake large-scale projects that lead to degradation of protected areas. This damage may be entirely unintended, stemming from a lack of appreciation of the value of the protected areas or of their location, or it may be a direct consequence of intentional resource extraction not subject to normal legal control. Although it is not always easy to distinguish between these two kinds of damaging activities, the difference is relevant to formulating the appropriate response. When lack of information is responsible, information transfer and support to build up the administrative and management capabilities is called for. Where elites operate illegally outside democratic control, a more activist response is appropriate. This distinction between incentives and disincentives is also not hard and fast; for instance, there may be a broad consensus in a country to utilize wildlife or convert wildlands with species extinction as a result. Nonetheless, the distinction remains useful in determining the relative emphasis on incentives or disincentives by the international community.

## Problems with Resource Pirates

Parks may be converted (e.g., to plantation agriculture) or their resources may be extracted (trees, minerals), with all the benefits going to powerful outsiders rather than society at large (Gillis and Repetto, 1988). This happens when governments, corporations, military leaders, or some combination of these components are able to ignore the laws of the land because they are immune to the retribution normally meted out to trespassers of the law. This resource theft by elites is all the more insidious because it undermines the credibility of any attempts to curb local encroachment.

Resource theft by the powerful reaches its most extreme form where political instability has culminated in civil war, armed insurgency, anarchy, or malignant dictatorship. Under those circumstances, the only external support for conservation that remains effective is that aimed directly at local communities, provided that the risk of having all protection undone overnight is acceptably low. Some countries are now in a state of social disintegration in which effective conservation has become impossible, and perhaps all bilateral and multilateral action from the outside should be abandoned. One practical lesson to be drawn from this is that international conservation assistance should not be limited to a single country within a biogeographic province.

The political situation in countries with self-serving elites makes change from the inside all but impossible. What, then, should be the response of the international community where biodiversity values are threatened by illegitimate governments or elites? Of course, the world is burdened with many illegitimate and suppressive regimes, and so far not surprisingly the response has been inspired mainly by the perceived strategic or economic interests of the dominant powers. However, the gradual strengthening of international institutions since the end of the Cold War gives hope for the future.

Direct political pressure from the outside concerning instances of threats to internationally known parks may improve protection in isolated cases. International activism may help to shame uncooperative governments into compliance with international norms or agreements. A mechanism for this is monitoring of the state of protected areas by an independent body with a reputation for veracity and widespread publication of the results. At the moment, there is no such independent body, whose role would resemble that of Amnesty International in the human-rights arena.

In the end, the most effective strategy for the outside world will be to encourage democratic elements and to stimulate education and freedom of information. Where citizens have the power to challenge illegal government actions, such challenges will eventually come to serve the common good. In particular, NGOs and citizens should have the right to challenge executive decisions in court to test the legality of the government's own actions. Accountability and transparency will also be promoted by the presence of active and unbridled news media. Thus, if our objective is to improve biodiversity conservation in other countries, one of the most effective actions may be to promote democracy, accountability, and transparency.

While democracy is necessary to foster biodiversity protection, it is not sufficient—economics and education are needed, too. First, some basic level of economic security must be reached so that the citizens become interested in ecological sustainability and long-term planing and subsequently in biodiversity protection (Southgate and Clark, 1993). The demand for biodiversity may behave like a luxury good, in that the value attached to it increases more than proportionately as incomes rise. One important implication is that international inequities in trade and access to markets that act as obstacles to economic development should be removed. Second, education is needed before people embrace intangible values such as the existence values of biodiversity. The notion that species have a right to exist regardless of their direct value to humans is not so widespread that its acceptance can be taken for granted.

## Problems with Intergovernmental Assistance

Sometimes, ill-conceived government projects unintentionally degrade protected areas. Government agencies may be unaware of the location of protected areas because good maps or interagency coordination are lacking, the agencies may not appreciate the true value of the resources residing in the protected areas and the (unpriced) services emanating from them, or they may not appreciate the impacts of intensive land uses at a park's boundaries or the latent benefits they provide. In other cases, competing interests predominate and overrule the conservation department. In these situations, the international community's best response is to draw these countries into binding treaties.

Several international treaties relevant to biodiversity conservation already exist, and some, such as the Convention on International Trade in Endangered Species (CITES), enjoy wide membership (WCMC, 1992). Susskind (1994) argues that they are also reasonably effective because governments realize it is in their own self-interest to comply rather than isolate themselves from the world community. Nonetheless, there is room for improvement in compliance with treaties concerned with biodiversity. For instance, the World Heritage Sites, covered by the World Heritage Convention, were not found to be in much better shape than the average protected rain forest area (see chapter 4).

In principle, the Convention on Biodiversity, arguably the most encompassing of all international green treaties adopted so far, has great potential to improve in situ protection of biodiversity, if the signatories agree to establish and maintain an international system of protected areas and set up an independent international monitoring organization. A monitoring body is needed because it can produce the reliable information required to identify the areas in which outside management support is needed, and because the public availability of information is often as effective as having active sanction mechanisms. Unfortunately, while the biodiversity convention explicitly recognizes the importance of in situ conservation, the discussion of its implementation has so far largely ignored the important role of protected areas (WWF, 1994). Although effective implementation is usually a painstakingly slow process, experience with previous negotiations has provided useful lessons on how to generate achievement of a treaty's objectives (French, 1992; Keohane et al., 1993). We cannot afford to miss this opportunity to strengthen the influence of the international community on the concern of individual nations for the biodiversity inside their territories.

## LAST STANDS: THE PLIGHT OF PROTECTED AREAS

It is clear, then, that tropical rain forests are in jeopardy—already greatly diminished in amount and distribution, and seriously threatened by a host of increasing pressures, most notably the actions of a rapidly growing human population worldwide that is laying claim to an ever larger share of land and resources. To help stem this destructive tide that is sweeping away unknown but certainly large numbers of species in tropical rain forests—as well as in other biomes—the world community has placed most of its efforts on establishing protected areas in which consumptive uses by humans are limited. Where protected areas have been properly planned, created, and managed, they have had success in meeting their goal of conserving biodiversity, and have provided a host of other important ecological benefits as well.

But these successes have been far too rare—protected areas are too few in number and often too small in size, and many of them are in poor condition, plagued by inadequate finances and only marginal local support. The resolve to uphold protected areas is often missing, because many individuals and groups, both private and public, often perceive it to be in their interest to exploit or convert the forest resource. Interest in safeguarding protected areas often derives primarily from other groups who have less immediate impact on the forests, including nations (although some have often wavered in their support), the international community, and future generations locally and worldwide. While the total global benefits of establishing and carefully managing protected areas may far outweigh the immediate gains obtained when a particular forest is exploited, it is often difficult to translate this generalized interest into immediate action on the local front.

We believe that the well-being of all parties—local and international, public and private, current and future—can best be served over the long term by the establishment and management of protected areas following the four principles outlined in this chapter. In summary, these principles are that (1) protected areas will always be in need of active defense, no matter how great their benefits are to local communities or to society at large; (2) beneficiaries of protected area services who now receive the benefits for free should be prepared to pay to support protected areas; (3) effective solutions require the involvement of all stakeholders, including representatives of both the local and the international community; and (4) while delegation of management to local communities is to be encouraged, there is always a role for the national government as the representative of the nation or the international community. Particular courses of action for establishing and managing protected areas will, of course, vary among locations; as we noted, careful analysis of the causes of protected area degradation must be done on a site-by-site basis, and solutions must be tailored to the site-specific needs. Still, the general framework described in this book should make such analyses easier and should help in designing fitting solutions.

The major challenge lies in getting started, in generating the resolve, as well as the financial support, necessary to shore up the network of protected areas upon which much of our collective hope for conserving biodiversity rests. Given the forces driving forest destruction, it is tempting to postpone action, in the hope that future circumstances will make the task easier. But every delay will be marked by further extinctions. The amount of tropical rain forests—and the host of species they contain—that will be protected for future generations depends on our actions within the remainder of this decade.

## NOTES

We have benefited from discussions with numerous people. Special thanks go to Keyt Fischer, Robert Healy, Nick Salafsky, Tom Struhsaker, and John Terborgh.

This work was supported by a cooperative agreement between the U.S. Agency for International Development and the Duke University Center for Tropical Conservation. Carel P. van Schaik's fieldwork is supported by the Wildlife Conservation Society.

## REFERENCES

Boo, E. 1990. *Ecotourism: the potentials and pitfalls.* Washington, D.C.: World Wildlife Fund.

Boza, M. A. 1993. Conservation in action: past, present, and future of the national park system of Costa Rica. *Conservation Biology* 7:239–47.

Brandon, K. 1993. *Bellagio conference on ecotourism: briefing book.* New York: Rockefeller Foundation.

Brechin, S. R., S. C. Surapaty, L. Heydir, and E. Roflin. 1994. Protected area deforestation in South Sumatra, Indonesia. *The George Wright Forum* 11:59–78.

Brown, M., and B. Wyckoff-Baird. 1992. *Designing conservation and development projects.* Washington, D.C.: Biodiversity Support Program.

Fearnside, P. M. 1993. Deforestation in Brazilian Amazonia: the effects of population and land tenure. *Ambio* 22:537–45.

French, H. F. 1992. *After the earth summit: the future of environmental governance.* Worldwatch Paper No. 107. Washington, D.C.: Worldwatch Institute.

Gillis, M., and R. Repetto, eds. 1988. *Public policy and the misuse of forest resources.* New York: Cambridge University Press.

Gorchov, D. L. 1994. Natural forest management of tropical rain forests: what will be the "nature" of the managed forest? In *Beyond preservation, restoring and inventing landscapes,* A. D. Baldwin Jr., J. DeLuce, and C. Pletch (eds.), 136–53. Minneapolis: University of Minnesota Press.

Healy, R. G. 1994. Merchandise as a means of generating local benefits from ecotourism. *Journal of Sustainable Tourism* 2:137–51.

Keohane, R. O., P. M. Haas, and M. A. Levy. 1993. The effectiveness of international environmental institutions. In *Institutions for the earth: sources of effective international environmental protection,* P. M. Haas, R. O. Keohane, and M. A. Levy (eds.), 3–24. Cambridge, Mass.: MIT Press.

Lynch, O. 1992. *Securing community-based tenurial rights in the tropical forests of Asia: an overview of current and prospective strategies.* Washington, D.C.: Center for International Development and Environment, World Resources Institute.

McNeely, J. A. 1993. Economic incentives for conserving biodiversity: lessons for Africa. *Ambio* 22:144–50.

Myers, N. 1993. Biodiversity and the precautionary principle. *Ambio* 22:74–79.

Neumann, R. P., and G. E. Machlis. 1989. Land-use and threats to parks in the neotropics. *Environmental Conservation* 16:13–18.

Oates, J. F. 1995. The dangers of conservation by rural development—a case-study from the forests of Nigeria. *Oryx* 29:115–22.

Panayotou, T., and P. S. Ashton. 1992. *Not by timber alone: economics and ecology for sustaining tropical forests.* Washington, D.C.: Island Press.

Redford, K. H. 1992. The empty forest. *BioScience* 42:412–22.

Skole, D. L., W. H. Chomentowski, W. A. Salas, and A. D. Nobre. 1994. Physical and human dimensions of deforestation in Amazonia. *BioScience* 44: 314–22.

Southgate, D., and H. L. Clark. 1993. Can conservation projects save biodiversity in South America? *Ambio* 22:163–66.

Susskind, L. E. 1994. *Environmental diplomacy: negotiating more effective global agreements.* New York: Oxford University Press.

Thiollay, J. M. 1992. Influence of selective logging on bird species diversity in a Guianan rainforest. *Conservation Biology* 6:47–63.

WCMC [World Conservation Monitoring Centre]. 1992. *Global biodiversity: status of the earth's living resources*. London: Chapman and Hall.

Wells, M., and K. Brandon, with L. Hannah. 1992. *People and parks: linking protected area management with local communities*. Washington, D.C.: World Bank, World Wildlife Fund, and U.S. Agency for International Development.

West, P. C., and S. R. Brechin, eds. 1991. *Resident peoples and national parks: social dilemmas and strategies in international conservation*. Tucson: University of Arizona Press.

WorldWide Fund for Nature. 1994. *The conference of the parties to the convention on biological diversity*. Position paper presented at the first meeting, WorldWide Fund for Nature, Nassau, the Bahamas, 28 November–9 December. Gland, Switzerland: WorldWide Fund for Nature.

# Index


Access rights, 134
Agency for International Development. *See* U.S. Agency for International Development (USAID)
Agenda 21, 59
Agriculture, 58, 183, 212
  commodity prices, world, 121
  encroachment on protected areas, 50, 79
  swidden. *See* Swidden agriculture
  technical and financial assistance for, 221, 222
Alienation, right to, 134
Amboseli National Park, 195
ANCON (National Association for the Conservation of Nature), 108
Angola, 16, 81
Annapurna Conservation Area, 94, 103
Arfak Mountains Nature Reserve, 52–55
Argentina, 70
Armed conflicts, 79, 225
  financed with valuable natural resources, 81
Australia, state of protected rain forests in, 74

Bali, 42
Bangladesh, 73
BAPPENAS, 48
Baumol, W. J., 195–96
Belize, 154–57
  Protected Areas Conservation Trust, 178
Benin, 37
Berkes, F., 102
Bhutan, 37
Bilateral and multilateral lending agencies, 136, 174, 218, 222
  *See also* Funding for biodiversity protection, international
Biodiversity
  definitions of, 3–4, 16–17, 116–20
  ecological foundations of. *See* Ecological foundations of biodiversity protection
  economic value of. *See* Economic value of biodiversity
  indigenous people and. *See* Indigenous people, biodiversity objectives versus needs of
  land-use cascade and, 29–30
  nonuse values of, 8, 163, 166, 168
  as political term, 116
  politics of. *See* Politics of biodiversity
  property rights and. *See* Property rights and biodiversity
  sustainable use and, 4, 6, 98–99
    questionable assumptions underlying. *See* Questionable assumptions underlying biodiversity policy
  in tropical rain forests, 9, 17, 36
  user rights and conservation of. *See* User rights and biodiversity conservation
Biodiversity Convention, 118, 121, 123, 125
*Biodiversity* (Wilson ed.), 117
Biosphere reserves, 94, 99
Biotechnology, 123–24
Bogor Botanic Gardens, 41–42

Bolivia, 69, 152–54, 176
National Fund for the Environment, 178
Borneo, 38, 39, 43, 44, 74
Botswana, 136
Brandon, Katrina, 90–111
Brazil, 79, 126, 223
extractive reserves in, 95
history of user rights in, 136
inequitable distribution of land in, 141
subsidy for livestock ranching, 97, 164–65
Brazilian Amazonia, 69–70
Brown, Katrina, 176
Brundtland Report, 95
Brunei, 37
Brussard, P. F., 117–18
Buffer zones, 94, 109, 221
management of, 109
production forests as, 57
Burley, F. William, 117
Burundi, 71
Bush, George, 123
Business interests, involvement in protected area management of, 55–56

Cambodia, 16, 42, 73
Cameroon, 71
CAMPFIRE, 52, 94
Carbon dioxide regulation, 149, 165, 214
*Caring for the Earth: A Strategy for Sustainable Living,* 5, 95, 98–99
Cassiavera, cultivation of, 49, 50–51
Central Africa
state of protected rain forests in, 70–71
Central African Republic, 37
Central America
inequitable distribution of land in, 139–41
state of protected rain forests in, 66–69
Cerro Campana National Park, 66, 77
Chimane Forest Reserve, 152–54
China, population pressures on forest reserves in, 141
CITES. *See* Convention on International Trade in Endangered Species (CITES)
Class-based inequities in user rights, 142
Climate change, 39, 212
extinctions due to, 19
Clinton, Bill, 123
Colombia, 70
indigenous communities in, 95
Colonialism and user rights, 135–36
Community Baboon Sanctuary, Belize, 154–57
Compensation, 187–203, 208, 220
criteria for judging the necessity of, 188–95
conflict mitigation, 194–95
efficiency, 189–92
equity, 192–93
ethics, 193
legality, 188–89
estimating the value of, 198–99
forms of, 188, 200
practical aspects of, 195–202
selection of recipients, 199–200
strategic behavior and, 195–98, 206
summary of positive and negative aspects of, 202–203
support for conservation, generating local, 200–202
Concession policies, 49, 80, 164
Conflict mitigation, compensation as form of, 194–95
Congo, 71
Conservation and development. *See* Development, conservation and
*Conservation Biology,* 117–18
Conservation International, 152, 176
Conservationists, interdisciplinary interaction between social scientists and, 107
Conservation Needs Assessment (CNA) project, Papua New Guinea, 40–41
Contingent-valuation method (CVM), 168–69, 170–71, 172
Convention on Biological Diversity, 59, 177, 227

Convention on International Trade in Endangered Species (CITES), 23, 121, 123, 226–27
Corbett National Park, 81
Corcovado National Park, 79
Corridors of natural habitat, 42, 212
production forests as, 57
Costa Rica, 141–42, 149–52, 176
allocation of land area for biodiversity, 46
deforestation rate, 20, 21
Council on Environmental Quality, U.S., 117
Craven, Ian, 52
Cutervo Park, 77

Davies, G., 41
Deacon, R. T., 83–84
Debt-for-nature swaps, 152, 175–76, 177, 178, 223
Deer, effect of extermination of predators of, 24–25
De facto rights, 134–35, 141–42, 144
Defense of reserves. *See* Management of reserves, defense and enforcement
De jure rights, 134–35, 141–42, 144
Democracy, fostering, 226
Development
conservation and, 5
fundamental conflict between, 7–8, 47, 58
questionable assumptions about results of poverty mitigation and, 104–106
*Development and the Environment* (World Bank), 97
Diversity of communities and biotic processes, 120
Diversity of higher taxonomic levels, 120
Dodo, 23–24
Doi Inthanon National Park, 80
Dominican Republic, 176
Dugelby, Barbara, 64–86
Dumoga-Bone National Park, 46–47
Dutch Association of Electricity-producing Companies Foundation, 56

Earth Summit. *See* United Nations Conference on Environment and Development (UNCED), 1992
Ecological fallacy applied to biodiversity. *See* Questionable assumptions underlying biodiversity policy
Ecological foundations of biodiversity protection, 36–60
determination of adequate protection, 37–38
future challenges, 58–59
industry's involvement in protected area management, 55–56
lessons learned, summary of, 60
local community involvement, 46–47, 51–55
lowland rain forests, 38–39
national and local development, integrating protected areas with, 46–47
outside protected areas, 56–58
planning a protected area network, 39–45
regional development, integrating protected areas with, 47–51
*Economics and Biodiversity* (IUCN), 100
Economic value of biodiversity, 8, 84–85, 115, 162–84
"free riders" problem, 162, 166
measurement of benefits and costs, 166–69
nonuse values. *See* Nonuse values of biodiversity
policies, markets, and market failures, 164–66
political considerations, 162–63
public-goods nature of forest outputs, 166
stakeholders, interaction of, 163
valuation techniques, 166–69
case studies of nonmarket methods, 169–74
Ecosystem diversity, definition of, 120
Ecotourism, 154–57, 162, 165, 167, 170, 171–72, 205, 220

Ecuador, 145–46, 176
Education, 204, 226
Efficiency of allocation of land for protection, 189–92
Egypt, 174
El Salvador, 79, 139–41
Employment opportunities, 220
Endangered Species Act, 16
Endemism, 39
Enforcement of rules of reserves. *See* Management of reserves, defense and enforcement
Environmental economics, 166–69
Equatorial Guinea, 71
Equity
  of compensation, 192–93
  in forest use and managements, 139–43
  vertical, 192, 206
Ethics
  of compensation, 193
  of denying the poor access to protected areas, 215
Ethiopia, 29, 148
Exclusion rights, 134
Exotic species, 25–26, 76, 212
Externalities, 187
  compensation of people subject to. *See* Compensation
  negative versus positive, 193–94
Extinction
  balanced by speciation, 17
  evolutionary process and, 16–17
  leading causes of, 15, 19–26
    fragmentation, 21–22
    habitat loss, 19–20
    interactions among, 26
    introduced species, 25–26
    overkill, 22–23
    secondary extinctions, 23–25
  minimizing species loss through protection. *See* Minimizing species loss through protection
  problem of estimating, 17–18
  proximate versus ultimate causes of, 19
  in tropical rain forests, 9, 15, 17
Extraction. *See specific forms of extractive activities*
Extractive reserves, 91, 93–94, 95
  questionable assumptions about management of, 101–103

Ferraro, Paul J., 187–209, 197, 198
Fires, 74, 212, 213
Fishing, 76, 213
Flexibility in conservation management, 59
Foreign debt, pressure to repay, 85
Foundation for the Philippine Environment, 178
Fragmentation, 41, 42
  distance between fragments, 42
  extinctions due to, 21–22
  land-use planning and, 27
  rule of thumb, 21
Framework Convention on Climate Change, 121
Funding for biodiversity protection, international, 46–47, 48, 109–10, 175–78, 215–16, 223–24

Gabon, 71, 85
GATT, 25, 124
GEF. *See* Global Environmental Facility (GEF)
Gender and user rights, 142–43
General Agreement on Tariffs and Trade (GATT), 25, 124
Genetic diversity, definition of, 119
Genetic materials contributing to commercial agricultural products, 162
Gentry, A. H., 18
Ghana, 72, 174
*Global 2000,* 117
Global benefits of protection of biodiversity. *See* International stakeholders
Global Biodiversity Strategy, 4
Global Environmental Facility (GEF), 48, 49–51, 124, 174, 176–77, 178, 222, 223
Global warming, 149, 214
Guards. *See* Park guards
Guatemala, 69, 79, 176
Gunung Leuser National Park, 80

Habitat loss
extinctions due to, 19–20
land-use planning and, 27
Haiti, 16, 29
Hardin, Garrett, 136–37
Hatam people, Irian Jaya, 53
Hedonic travel-cost method, 167
Highways of Penetration project, Peru, 27
Horwich, Robert, 154
Household discount rates, 204
Hunting, 76, 122–23, 212, 213
by indigenous people, 54
overkill, 22–23
Hydroelectric dams, 80, 213

ICDPs. *See* Integrated conservation-development projects (ICDPs)
Iguazu National Park, 70
Immigration, 206, 222–23
compensation and, 196
conflict for user rights and, 144–46
land titling systems and, 197
*See also* Transmigration
INBio, 149–52
Incentives and disincentives
compensation of local communities. *See* Compensation
for degradation of natural preserves, 81–86, 208
to promote conservation, 164–65, 203–209, 220–21
advantages of positive economic incentives, 206–207
affecting resident behavior, 203–206
finding the linkages, 207–209
questionable assumptions, 100–101
recommendations, 208
India
history of user rights in, 135–36
state of protected rain forests in, 72–73
Indigenous people
biodiversity objectives versus needs of, 7, 70
conflicts over resource use, 106
locally managed reserves. *See* Locally managed reserves
myths concerning use of natural environment by, 51–52
political movement to protect the rights of, 4, 6–7
questionable assumptions about, 106
rapid population growth of, 80
as threat to rain forest parks, 79–80, 217–18
dealing with, 218–24
user rights issues. *See* User rights and biodiversity conservation
*See also* Local communities
Indigenous reserves, 91, 94–96
questionable assumptions about management of, 101–103
Indochina, state of protected rain forests in, 73–74
Indonesia, 46–47, 79, 84
deforestation of, 36
National Conservation Plan, 39, 42, 43, 74
production forests of, 56–57
protected areas in, 37
state of protected rain forests in, 74
transmigration project, 27, 74, 136, 145
*See also individual islands*
Indonesian Ministry of Forestry, 49–50
Industrialization, 222
Industry's involvement in protected area management, 55–56
Institutions as threat to protected areas, 80, 213, 218
dealing with, 224–27
Integrated conservation-development projects (ICDPs), 47–51, 91, 93, 99
core protected area in, 93
defining factor of, 91, 92, 93
examples of, 94
failure to achieve their objectives, 105
funding for, 109–10
problems with, 108–10
Intellectual property rights, 123–24

Inter-American Foundation, 155, 156
International agencies
as threat to rain forest parks, 80, 218
dealing with, 226–27
International Environmental Protection Act, 117
International stakeholders
conflicts with local and state interests, 148–49
funding of biodiversity protection. *See* Funding for biodiversity protection, international
global benefits of protection of tropical rain forests, 165–66, 214
International Tropical Timber Organization, 152–53
International Whaling Convention, 123
Introduced species, 25–26, 76, 212
Irian Jaya, Indonesia, 52–55, 74
"Islands" of habitat, 21
biogeography of, 41

Jamaica, 176
Java, 39, 41, 42

Kaltim Primacoal (KPC), 56
Kanha National Park, 81
Kanum tribe, Irian Jaya, 54
Kayapo Indians of eastern Amazonia, 52, 79–80
Kenya, 102
Keohane, R. O., 216
Kerinci-Seblat National Park (KSNP), 47–51, 77
Khao Yai park, 74
Knetsch, J. L., 193, 198
Korup National Park, 71
Kothari, A., 73
Kramer, Randall A., 3–11, 162–84, 187–209, 198, 207, 212–28
Kramer R. A., 199
Kruger National Park, 194
Kubu people of Sumatra, 51
Kutai National Park, 38, 55–56, 74, 77

Labor devoted to illegal activities, incentives and disincentives to change, 204–205
Labor productivity, improving, 204
Lacandon Biosphere Reserve, 80
Laguna del Tigre National Park, 69
Lake Victoria, 42, 128
Land-use cascade, 27–30, 221
degraded land, 29
conversion to productive land, 221–22
extensively used land, 28, 30–31
intensively used land, 29
wildlands, 28
Land-use planning, 32, 212
protected areas and, 27–30
Laos, 42, 73
LaPalme, Sharon, 133–59, 217
Large-scale planning units, 94
Legal justification for compensation, 188–89
Liberia, 81
Local communities
compensation of. *See* Compensation
degradation of protected areas by, 217–18
dealing with, 218–24
incentives and disincentives:
for degradation of resources, 82–85, 208
to promote conservation. *See* Incentives and disincentives, to promote conservation
involvement in protected areas management, 46–47, 51–55, 101–103, 133, 217
as threat to rain forest parks, 78–79, 133
user rights issues. *See* User rights and biodiversity conservation
*See also* Indigenous people
Locally managed reserves, 91, 93–94, 95
examples of, 91
human objectives of, 91, 92
methods of enhancing, 110–11
technical assistance, support, and training for, 110–11

Logging, 58, 74, 147, 213
concessions, 49, 80
conservation of biodiversity in production forests, 56–58
illegal, 76
Lovejoy, Thomas, 117
Lowland rain forests, 38–39, 50

MacKinnon, Kathy, 36–60
McNeely, J., 100
Madagascar, 29, 79, 169–72, 176
National Environmental Action Plan, 174–75
state of protected rain forests in, 72
Maderas Industrializados de Quintana Roo (MIQRO), 147
Malawi, 71
Malaysia, 74
Mammals, fragmentation's effect on, 22
Management of reserves
defense and enforcement, 69–70, 84, 108, 194, 204–205, 215
foreign involvement in, justification of, 216–17
local communities, involvement of, 46–47, 51–55, 101–103, 133, 217
new protection paradigm, 212–28
actors behind protected-area degradation, 217–18
basic principles, 214–17, 228
big players, dealing with threats posed by, 224–27
last stands, 227–28
small players, dealing with threats posed by, 218–24
questionable assumptions about, 101–103
Management rights, 134
Manas National Park, 81
Mantadia National Park, 169–72
Manu National Park, 80
Marind tribe, Irian Jaya, 43
Marori tribe, Irian Jaya, 54
Maya Biosphere Reserve, 69
Maya of Quintana Roo, 147
Medicines, traditional, 162
Merck International, 150
Mexico, 147, 176
state of protected rain forests in, 66–69
Meyer, P. A., 198
Migrants. *See* Immigration; Transmigration
Migration policies, 222–223
Military role in protection of parks, 224
Minimizing species loss through protection, 15–33
comprehensive land-use planning and protected areas, 27–30
estimation of species loss, 17–19
large protected areas, 15, 25, 31–32
leading causes of extinction. *See* Extinction, leading causes of
population growth pressures, 15–16
Minimum viable population, area required to protect, 41
Mining, 50, 55–56, 80, 213
*See also* Extractive reserves
Miranda, Marie Lynn, 133–59, 217
Montreal Protocol, 121
Multilateral and bilateral lending agencies, 136, 174, 218, 222
Multiple-use areas, 94
Murphee, Marshall, 102–103
Myanmar, 42, 73, 79, 81

NAFTA, 121
Nair, S. C., 73
National Environmental Action Plans (NEAPs), 27, 174–75, 180
National governments
support for protected areas, 46
as threat to protected areas, 218
dealing with, 224–27
National Institute for Biodiversity (INBio), 149–52
National parks
European concept of, 5
first in United States, 5
*See also* Parks; Protected areas
National Park Service, U.S.
establishment of, 5
National Park System, U.S., 64

Native Lands, 110
Natural and social history of the landscape, 128–29
Nature Conservancy, The, 110, 117
Nepal, 144
NGOs. *See* Nongovernmental agencies (NGOs)
Niger, 140
Nigeria, 71
Nongovernmental agencies (NGOs), 178, 223, 224
  environmental, 97–98, 110
  management of parks by, 102
Nonuse values of biodiversity, 8, 163, 166, 168
  existence value, 163
  passive-use value, 163
Norse, E. A., 117
North American Free Trade Agreement (NAFTA), 121
Norway, 192

Oates, W. E., 195–96
Office of Technology Assessment (OTA), 117
Operation CAMPFIRE, 52, 94
Opportunity-cost analysis, 170–71
Opportunity costs, 214, 220
  compensation for loss of. *See* Compensation
  defined, 188
  of investing labor in destructive activities, 204
Organic Act, 5
Overkill, 22–23, 27
Ownership rights. *See* Property rights and biodiversity
Ozone layer, greenhouse gases affecting the, 216, 217

Panama, 66, 108
Papua New Guinea, 40–41, 42, 74
Park guards, 69–70, 84, 194
  low salaries and corruption, 224
Parks
  accommodation of human needs, effect of, 32
  degradation of, 92–93
  elimination of uses that conflict with biodiversity conservation, 108
  enabling legislation, 32
  enforcement, need for improved, 108
  maintenance of, 107
  new, 32, 108
  objectives of, 92, 107–108
  redesigning of, 212
  *See also* Protected areas
Payne, J., 41
Peres, C. A., 70
Peru, 70, 79, 84
Pharmaceuticals, 162, 165
Philippines, 74, 176
Pico da Neblina National Park, 70
Pirates, resource, 81, 85–86, 225–26
  *See also* Trade in wildlife products, international
Planning stages of conservation projects, 59
Poaching. *See* Trade in wildlife products, international
Poland, 176
Political instability
  deforestation and, 83–84
  resource theft and, 225
Politics of biodiversity, 120–21
  relationship of biodiversity conservation to property, 125–28
  short-term coalitions, 121
Polonoreste Project, Brazil, 27, 136
Population growth
  land-use cascade and, 30
  local management practices and, 102
  pressure of, 15–16, 30, 77, 137–39, 140–41, 158, 183
Potential Pareto-improvement (PPI), 189
Poverty, 8, 201
  protection of parks and, 84
  questionable assumptions about results of mitigation of, 104–106
Predators, effect on prey species of extermination of, 24–25

Production forests, conservation of biodiversity in, 56–58
Property rights and biodiversity, 121–24, 125–28
  intellectual property rights, 123–24
  landed property rights, 122
  wildlife property rights, 122–23
Protected area degradation
  large players, 218
    dealing with threats posed by, 224–27
  small players, 217–18
    dealing with threats posed by, 218–24
Protected areas
  area covered by, 212
  changing views of, 4, 5–6
  compensation to reduce pressure on. *See* Compensation
  ecological foundations of biodiversity protection. *See* Ecological foundations of biodiversity protection
  economic decision to devote funds to, 165
  incentives and disincentives to defend. *See* Incentives and disincentives
  minimizing species loss through. *See* Minimizing species loss through protection
  new protection paradigm for. *See* Management of reserves, new protection paradigm
  objectives of, 92
  state of rain forest preserves. *See* State of protected rain forests
  *See also* Parks

Questionable assumptions underlying biodiversity policy, 90–91, 95–106
  ecological fallacy, 96
  implications for biodiversity conservation, 107–11
    integrated conservation-development projects, 108–10
    locally managed reserves, 110–11
    parks, 107–108
  incentives, 100–101
  management, 101–103
  method, 96–98
  poverty mitigation and development, 104–106
  social, 106
  technology, 103–104
  use, 98–99

Random-utility model, 167–68
Range loss and habitat protection for Southeast Asia primates, 44, 45
Ranomafana National Park, 197
Recreation-demand model, 170, 171–72
Redford, Kent H., 106, 110, 115–29
Regional Conservation Area System, Costa Rica, 94
Regional impact assessment (RIA), 50
Regional support for protected areas, 46, 47–51
Research and action agenda, recommended, 181–83
Resource pirates, 81, 85–86, 225–26
Rhinoceroses, 23, 76
Roads, 85, 213, 222–23
  negative impacts of, 47, 50, 75–76, 80
Robinson, J., 110
Royal Chitwan National Park, 194
Royal Society of the United Kingdom, 57
Rwanda, 16, 71, 79

Sabah, East Malaysia, 57
Sanchez, P. A., 104
Sanderson, Steven E., 115–29
Sapo National Park, 81
Scale of biodiversity conservation, 124–25
Secondary extinctions, 21, 23–25
  land-use planning and, 27
Senegal, 37
Serrania de Macarena, 77
Seymour, F., 97
Sharma, Narendra, 162–84, 214
Sharma, U. R., 194
Shaw, W. W., 194

Shenandoah National Park, 77
Shyamsundar, Priya, 199
Sierra Leone, 81
Slash-and-burn agriculture. *See* Swidden agriculture
Smith, V. K., 167
Social and natural history of the landscape, 128–29
Social scientists, interdisciplinary interaction between conservationists and, 107
Somalia, 16
South America
  inequitable distribution of land in, 139–41
  state of protected rain forests in, 69–70
South Asia, state of protected rain forests in, 72–73
Southeast Asia, state of protected rain forests in, 74
Species diversity, definition of, 119–20
Sri Lanka, 73
  National Environmental Action Plan, 174, 175
Staffing of parks, 49, 69–70, 84
  *See also* Park guards
State of protected rain forests, 64–86
  overview, 64–65
  perpetrators of threats, 217–18
    indigenous people, 79–80, 217–18
    large institutions, 80, 218
    local residents, 78–79, 217–18
    resource pirates, 81, 225–26
  regional overviews, 65–77
    Australia and Papua New Guinea, 74
    Central Africa, 70–71
    Central American and Mexico, 66–69
    current threats to individual countries, 68
    data compilation procedure, example of, 66
    Indochina, 73–74
    Madagascar, 72
    quantitative comparisons, 75–77
    South America, 69–70
    South Asia, 72–73
    Southeast Asia, 74
    West Africa, 71–72
  roots causes of the crisis, 81–86
    individuals, pressure from, 82–85
    official projects and organization resource theft, pressure from, 85–86
  system of categories of protected areas, 65
Strategic behavior, compensation and, 195–98, 206
Subsistence use of resources, 213
Sumatra, 39
  deforestation of, 36–37
Sustainable use, 223
  biodiversity and, 4, 6, 98–99
    questionable assumptions underlying. *See* Questionable assumptions underlying biodiversity policy
  objectives of sustainable development, 5–6
  strategy for, 180–81
Swidden agriculture, 20, 47, 82–83, 104, 145, 205

Taï National Park, 72, 77
Takings, 188
Taman Negara conservation area, West Malaysia, 38
Tanzania, 37
*Technologies to Maintain Biological Diversity* (Office of Technology Assessment), 117
Technology
  adoption of, 104
  questionable assumptions about resource management capabilities of, 103–104
  technical assistance, support, and training for indigenous groups, 110–11
Terborgh, John, 15–33, 64–86
Thailand, 85
  deforestation of, 42–43, 73, 74
  usufruct rights in, 142
Thiesenhusen, William C., 139

Title systems, property, 142, 146, 197
Trade in wildlife products, international, 23, 76–77, 215, 224
Tradeoffs in protecting biodiversity, 178–79
"Tragedy of the commons," 136–37
Transmigration
  programs, 80
    Indonesia, 27, 74, 136, 145
Travel-cost method of valuing biodiversity, 167, 168
Tropical rain forests
  biodiversity in, 9, 17, 36, 213
  large-scale exploitation of, 10
  rate of destruction of, 36–37
  special attention warranted by, 8–10, 213
  the state of natural preserves. *See* State of protected rain forests
Troup, R. S., 135
Truck farmers, 79
Trust funds for protected areas, 177–78, 223

Uganda, 71
UNITA (National Union for the Total Independence of Angola), 81
United Nations Conference on Environment and Development (UNCED), 1992, 115, 125, 177
United Nations Conference on the Human Environment, 1972, 115
United Nations Development Fund (UNDP), 176
United Nations Educational, Scientific and Cultural Organization (UNESCO)
  Man and Biosphere Program, 5
United Nations Environmental Programme (UNEP), 98, 176
U.S. Agency for International Development (USAID), 6, 80, 117, 118, 152
  Tropical Forests and Biodiversity Support Program, 6
United States residents, value placed on tropical rain forest protection by, 172–74
Urbanization, 222
USAID. *See* U.S. Agency for International Development (USAID)
User rights and biodiversity conservation, 133–59
  case studies, 149–57
  common conflicts, 143–49
    case studies, 149–57
    commercial enterprises versus traditional users, 146–48
    local and state interests versus international stakeholders, 148–49
    states versus traditional users, 143–44
    traditional users versus new migrants, 144–46
  defining, 134–37
  devolution of, 133–34, 158
  future path for, 157–59
  historical background, 135–37
  issues shaping the devolution debate, 137–43
    equity, 139–43
    population pressure, 137–39, 140–41
  types of rights, 134–35
Usufruct rights, 122, 142

Valuation of biodiversity. *See* Economic value of biodiversity
van Schaik, Carel P., 3–11, 15–33, 64–86, 212–28
Vietnam, 73–74
Volan Colima National Park, 66

Wasur National Park, 54–55
Wells, M., 109
West Africa, state of protected rain forests in, 71–72
Wetlands, 58
Wildlife Conservation Society, 6, 224
Willingness to pay (WTP), 167
Wilson, E. O., 117

Withdrawal rights, 134
Women, user rights of, 142–43
World Bank, 48, 97, 110
  funding of conservation and development, 46–47, 176
  National Environmental Action Plans, 27
World Conservation Strategy, 5, 7, 8, 96
World Conservation Union (IUCN), 37, 75, 98, 100, 101
  system of categories for protected areas, 65, 92
World Heritage Convention, 227
World Heritage Sites, 75, 227
World Parks Congress
  1982, 93, 96
  1992, 95
World Wildlife Fund, 53, 54, 98, 155, 166, 176

Yasuni National Park, 69
Yayasan Sabah, 57
Yei tribe, Irian Jaya, 54
Yellowstone National Park, 5

Zaire, 71, 81
Zambia, 176
Zimbabwe, 37